PROJECT MANAGEMENT WORKBOOK AND PMP®/CAPM® EXAM STUDY GUIDE

PROJECT MANAGEMENT WORKBOOK AND PMP®/CAPM® EXAM STUDY GUIDE

Tenth Edition

Harold Kerzner, Ph.D.

Frank P. Saladis, PMP

WILEY

John Wiley & Sons, Inc.

Library of Congress Cataloging-in-Publication Data:

ISBN: 978-0-470-27872-7

Printed in the United States of America

10 9 8 7 6 5 4 3 2 1

Contents

Preface

The purpose of this workbook is to provide students of project management with meaningful exercises and homework problems that will enhance the knowledge of the subjects included in the textbook ***Project Management: A Systems Approach to Planning, Scheduling, and Controlling, Tenth Edition*** by Harold Kerzner, Ph.D., John Wiley, 2009.

The material in the workbook is directly related to the subjects and knowledge areas associated with the Project Management Institute® Project Management Professional PMP® Exam and the Certified Associate Project Manager CAPM® Exam and will provide a sound framework for exam preparation.

The workbook is designed to engage the student in activities that will provide practical application of the concepts of project management as described in the textbook and in the *PMI® Guide to the Project Management Body of Knowledge* (*PMBOK® Guide*) –Fourth edition.

Included in the workbook are topic specific glossaries, common project terms and acronyms, knowledge area summaries, examples of typical project management mathematic formulas and equations, key project terms and enjoyable crossword puzzles. The workbook also includes PMP® exam study tips, situational exercises, and sample questions designed to simulate the type of questions that may be encountered on the actual PMP® exam.

We hope you will find this book enjoyable and educational.

PROJECT MANAGEMENT WORKBOOK AND PMP®/CAPM® EXAM STUDY GUIDE

Chapter One

PROJECT MANAGEMENT OVERVIEW

Project management itself is not a new concept. It has been practiced for hundreds, even thousands of years. Any large undertaking requires a set of objectives, a plan (and continuous planning), coordination, the management of resources, and the ability to manage change. Today, the project management approach has become more formal and many organizations have adopted a "management by projects" approach. Some organizations have established project management offices or PMOs to assist them in developing standards for managing projects. As the project management discipline evolves, organizations around the world are experiencing the benefits of project management. These benefits include better scheduling of resources, improved estimating, higher levels of quality, early identification of issues and problems, and more effective measurement processes to assess success.

Projects are defined as temporary, they have a start and an end date, and they provide a unique product or service. Project management is the application of knowledge skills, tools, and techniques to project activities to meet project objectives. Effective project management is accomplished through the application and integration of project management processes that will assist in the initiation, planning, execution, monitoring, controlling, and closing of a project and each phase of a project. A project manager is assigned to a project and becomes accountable for the success of the project through effective management techniques, coordination between functional organizations, and the ability to apply the appropriate amount of managerial and cross-organizational support and guidance as the project is executed.

Glossary of terms   Key terms and definitions to review and remember

Deliverable   A tangible, verifiable work output.

Functional Manager   Generally, the manager who owns the resources that will be assigned to project activities. Functional managers are considered to be the technical experts and usually provide information about task duration and cost estimates. Project managers engage the assistance of functional managers (also known as line managers) to develop the project plan.

Non-Project Driven   Generally, these organizations do not have a project methodology in place and are arranged in a functional organizational structure. Work is associated with manufacturing and production lines. Projects are established as needed to improve or support functional lines and activities.

Project   A temporary endeavor undertaken to create a unique product, service, or result. Has a specific objective, defined start and end dates, funding limitations, consumes resources (human, equipment, materials) and is generally multifunctional or cross-organizational in nature.

Project Driven Organization   Also known as "project based." In these organizations all work is characterized through projects. Projects are arranged as separate cost centers and the sum of all project work is associated with organizational goals and strategic objectives.

Project Management   Application of knowledge skills, tools, and techniques to project activities to meet project requirements. Involves initiation, project planning, executing, monitoring, controlling, and closing of project phases and the total project.

Project Sponsor   The person or organization that authorizes the project and provides the financial resources.

Triple Constraint   A framework for evaluating competing project demands of Time (schedule), Cost (budget), and Scope (specifications) usually depicted as a triangle. Quality is commonly used in place of, or in addition to, Scope.

Activities, Questions, and Exercises

Refer to Chapter One of *Project Management: A Systems Approach to Planning, Scheduling, and Controlling* (10th Edition) for supporting information. Review each of the following questions or exercises and provide the answers in the space provided.

Dr. Kerzner's 16 Points to Project Management Maturity.

1. Adopt a project management methodology and use it consistently

2. Implement a philosophy that drives the company toward project management maturity and communicate it to everyone

3. Commit to developing effective plans at the beginning of each project

4. Minimize scope changes by committing to realistic objectives

5. Recognize that cost and schedule management are inseparable

6. Select the right person as the project manager

7. Provide executives with project sponsor information, not project management information

8. Strengthen involvement and support of line management

9. Focus on deliverables rather than resources

10. Cultivate effective communications, cooperation, and trust to achieve rapid project management maturity

11. Share recognition for project success with the entire project team and line management

12. Eliminate nonproductive meetings

13. Focus on identifying and solving problems early, quickly, and cost effectively

14. Measure progress periodically

15. Use project management software as a tool – not as a substitute for effective planning or interpersonal skills

16. Institute an all-employee training program with periodic updates based upon documented lessons learned

This exercise is intended to provide you with a basis and understanding of the major goals of an enterprisewide project management methodology and process for improvement. The 16 Points to Project Management Maturity are designed to assist an organization in achieving continuously higher levels of project performance by providing a baseline for assessing the current level of project management maturity and then developing steps to enhance existing processes and/or create new processes that will improve overall project performance.

Exercise: Review Dr. Kerzner's 16 points to project management maturity and identify the specific benefits associated with each point. Identify actions that may be taken to introduce, implement or further enhance the value of each of the listed points in an organization. .

Example:

1. Adopt a project management methodology and use it consistently.

Action: Provide management with supporting information about how project management can assist in achieving organizational objectives. Obtain best practices documentation from companies that are actively using project management processes and methodologies and provide a summary to executive management.

2. Implement a philosophy that drives the company toward project management maturity and communicate it to everyone.

_____

3. Commit to developing effective plans at the beginning of each project.

_____

4. Minimize scope changes by committing to realistic objectives.

_____

5. Recognize that cost and schedule management are inseparable.

_____

6. Select the right person as the project manager.

_____

7. Provide executives with project sponsor information, not project management information.

_____

8. Strengthen involvement and support of the line management.

_____

9. Focus on deliverables rather than resources.

_____

10. Cultivate effective communications, cooperation, and trust to achieve rapid project management authority.

_____

11. Share recognition for project success with the entire project team and line management.

_____

12. Eliminate nonproductive meetings.

_____

13. Focus on identifying and solving problems early, quickly, and cost effectively.

_____

14. Measure progress periodically.

_____

15. Use project management software as a tool—not as a substitute for effective planning or interpersonal skills.

_____

16. Institute an all-employee training program with periodic updates based upon documented lessons learned.

_____

2. Describe how project management may benefit an organization, impact organizational success, and assist in the achievement of strategic objectives.

_____

_____

_____

3. What are the three key factors that are commonly used to indicate project success?

   1. _____

   2. _____

   3. _____

4. What additional success factors could be considered to more effectively indicate successful completion of a project?

   1. _____

   2. _____

   3. _____

   4. _____

   5. _____

   6. _____

   7. _____

   8. _____

5. In many organizations the organizational structure itself may create management gaps, functional gaps, and operational islands that developed over time. These gaps may result in miscommunications and lower productivity. Describe some of the causes of these gaps and how the gaps can be effectively minimized.

   _____

   _____

   _____

6. Describe the term *stakeholder* and provide examples of the stakeholders associated with projects you are engaged in.

   _____

   _____

   _____

7. This diagram is commonly used to illustrate the relationship of the key elements of project success. Correctly label each side of the diagram (Figure 1-2).

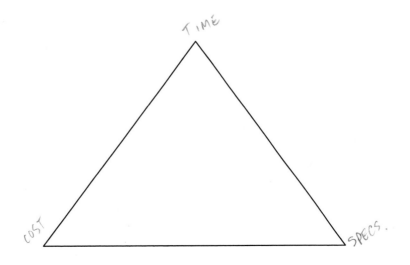

8. This diagram is referred to by project managers and other stakeholders as the _____.

   What is the significance of the diagram as it relates to the competing demands of a project?

   _____

   _____

   _____

9. Referring to Figure 1-2, describe at least three additional factors that may influence a customer's perception of project success.

   _____

   _____

   _____

10. Explain why establishing a good daily working relationship with functional managers and/or line managers is important to project success and is a critical responsibility of the project manager.

    _____

    _____

    _____

11. Explain the term *integration* as it relates to project management and describe the major roles and responsibilities of the project manager.

    _____

    _____

    _____

12. Define the roles of the functional manager and describe at least three challenges that a functional manager may encounter in an organization that engages in the management of multiple projects.

    _____

    _____

    _____

13. How can a project manager ensure that he or she establishes and maintains an effective and collaborative relationship with the project sponsor or project executive?

    _____

    _____

    _____

14. **Causes and effects.** Although all projects are unique, there are many common issues that are experienced by project managers and teams. Referring to the list of causes and effects, match two causes (only 2) to each effect and explain why the two were chosen.

    Causes:

    a) Top management does not recognize the activity as a project
    b) Too many projects going on at the same time

c) Impossible schedule commitments
d) No functional input into the planning phase
e) No one person responsible for the total project
f) Poor control of design changes
g) Poor control of customer changes
h) Poor understanding of the project manager's job
i) Wrong person assigned as project manager
j) No integrated planning and control
k) Company resources are overcommitted
l) Unrealistic planning and scheduling
m) No project cost accounting ability
n) Conflicting project priorities
o) Poorly organized project office

Effect		Explanation
1. Late completion of activities	Cause #1 i	
	Cause #2 j	
2. Cost overruns	Cause #1	
	Cause #2	
3. Substandard performance	Cause #1	
	Cause #2	
4. High turnover in project staff	Cause #1	
	Cause #2	
5. High turnover in functional staff	Cause #1	

Effect		Explanation
	Cause #2	
6. Two functional departments performing the same activities on one project	Cause #1	
	Cause #2	

15. Describe the major roles and responsibilities of the project manager:

16. Project managers are often challenged to influence functional managers who may have multiple projects to deal with and may be forced to compete with other project managers in the same organization for resources. Explain how the project manager can improve relationships with functional managers and influence them to provide the necessary resources to achieve the project manager's objectives.

17. The relationship between the project manager and the project sponsor or executive is a critical factor and can mean the difference between project success and failure. What actions can be taken by the project manager to ensure that a strong and supportive relationship exists between the project manager and the project sponsor?

18. According to the *PMBOK® Guide*—Fourth edition, a project is defined as:

19. The 42 project management processes described in the *PMBOK® Guide*—Fourth edition comprise the 5 major process groups. The 5 major process groups are:

20. In addition to the three elements of the triple constraint – Time, Cost, and Scope, what additional competing constraints may be encountered during project planning and execution?

Kerzner "Quick tips" for the Project Management Institute PMP® and CAPM® EXAM

The information in Chapter One is most closely related to the following topics in the *PMBOK® Guide*: THE PROJECT MANAGEMENT FRAMEWORK, PROJECT MANAGEMENT KNOWLEDGE AREAS—PROJECT INTEGRATION MANAGEMENT

An important item to remember is the Project Management Framework, as described in the *PMBOK® Guide*, which defines a total of 42 project processes that describe the activities generally found throughout a project's life cycle. These processes are organized into nine knowledge areas and represent five process groups: Initiating, Planning, Executing, Monitoring, and Closing.

The 9 knowledge areas of Project Management are: ***Integration management, Scope management, Time management, Cost management, Risk management, Human Resources management, Quality management, Procurement management, and Communication management.*** These knowledge areas and the subprocesses associated with them are connected through the "system" of project management and are all *Integrated, Interrelated,* and *Interdependent.* There are no independent components of the total project and all knowledge areas within the *PMBOK®* Guide may, in some way, impact any of the other knowledge areas. They are part of the total system of project management.

Important terms to remember

Change Control Board   A team or group designated or empowered to review and determine the value of a change and to approve or deny change requests.

Close Project   Utilizing the project management methodology, Project Management Information System, and expert judgment to complete the project and perform all final administrative procedures. The processes required to obtain formal acceptance and completion of project files for historical information.

Configuration Management   Process that will ensure that configuration changes (changes to features, functions, dimensions) are managed and approved to prevent or reduce the risks of additional cost, scope changes, or other impacts to the project.

Constraints and Assumptions   What are the limitations or boundaries you must operate within? What assumptions can be made about the project you have been assigned to manage? Assumptions in this context are items we can believe to be true, real, or certain.

Enterprise Environmental Factors   Items such as the organizational culture, industry standards that are in place, personnel administration guidelines, databases, and existing infrastructure.

Historical Records   Learn from past projects. Use lessons learned and experience. Also associated with professional and social responsibility.

Integrated Change Control   The 9 knowledge areas are managed in an integrated manner with an understanding that a change in one area can impact any or all of the other knowledge areas. Consider the impact of the change before implementing the change. Determine that a change has occurred through comparison of the baseline with actual results. Determine when to make a change and how to introduce the change to minimize the impact on ongoing operations.

Monitor and Control Project Work   Processes for managing work performance, managing change requests, utilization of earned value techniques, identifying corrective and preventive actions.

Organizational Process Assets   Standard policies such as safety, quality assurance, health, available planning templates, financial controls, change control procedures, and risk management processes.

Organizational Strategies   Consider the goals and objectives of your organization and how your project impacts or supports them. Make sure you can link your project to the organizational goals.

Preliminary Project Scope Statement   Describes the project and the desired objectives. The preliminary scope statement includes the definition of the project, the products and services to be delivered, major milestones, and acceptance criteria.

Project Charter   Authorizes the project and the use of resources. Assignment of the project manager is also included in a project charter.

Project Management Information System   Any system or group of systems working together to gather, store, and distribute information about your project. Examples: Time-reporting system, Accounting System, Project Software.

Project Management Plan   All of the actions necessary to integrate and coordinate the entire project effort including any subsidiary plans that have been established by the project team.

Project Plan   The approved document that provides the baseline for executing and managing the project.

SMART Objectives – Specific, Measurable, Attainable, Realistic, Time bound

Stakeholders   Consider who the key stakeholders are and also other stakeholders who may view your project as a threat or an obstacle to their projects. Determine who the negative stakeholders are and what risks they may introduce to the successful completion of your project.

Subsidiary Plans   Plans created to support the higher-level project plan. An example of a subsidiary plan is the Change Control Plan—the control processes in place to manage other knowledge areas such as Scope Change Control, Schedule Change Control, Cost Change Control, etc. The total or *integrated project plan* may include several subsidiary plans depending on the complexity of the project.

Integrated planning includes many processes, so be prepared to answer questions that may include several different processes related to a project situation. Become familiar with all process groups, and make sure you know the inputs, tools, techniques, and outputs of each process.

PMI® heavily emphasizes the importance of planning. Proper planning requires effective communication among the team and sound leadership from the project manager. The result is a project team that is more completely informed and has an understanding of the larger, integrated view of the project.

Additional tips and practice items for the PMP® exam are included in each chapter and in the section of the workbook entitled *PMP® Exam and PMBOK Guide® Review.*

Answers to Questions and Exercises

1.   16 Points

   1. Provide management with supporting information about how project management can assist in achieving organizational objectives. Obtain best practices documentation from companies that are actively using project management processes and methodologies and provide a summary to executive management.

   2. Identify and communicate the benefits of project management.

3. Establish a project kickoff process and project-planning methodology.

4. Set objectives clearly using SMART criteria.

5. Establish a performance measurement system using earned value management.

6. Establish guidelines and criteria for selection of a project manager. Emphasize soft skills as well as managerial skills.

7. Establish expectations with executives at project start-up.

8. Communicate project sponsor support and executive support to the team. Understand line manager priorities. Create a positive working relationship.

9. Prepare and communicate acceptance criteria. Communicate the scope statement.

10. Obtain sponsor and executive support, establish clear objectives, develop a communications plan.

11. Reward and recognize project teams, develop team building activities.

12. Create meeting guidelines. Meet only when necessary. Define the meeting purpose, create an agenda, and manage time effectively.

13. Develop a risk management plan and a process for managing issues.

14. Use earned value management and establish success metrics. Conduct reviews after each project phase.

15. Identify a software application that will be accepted and used by project managers. Provide the appropriate training.

16. Establish a Project Management Office, require documentation of lessons learned, and ensure that management support is visible.

2. Control of changes, consistent approach, improve quality, reduces risk, improves estimating ability.

3. On schedule, within budget, within performance specifications (and quality requirements).

4. Customer satisfaction. Add on business, employee satisfaction, no disruption of operations, minimal changes to the scope of work, executive management recognition of the project team, minimal conflicts among team members and organization units, fully operational and accepted product or service deliverables.

5. Functional units may develop their own culture, management hierarchy may affect the ability to communicate, protection of area of responsibility (turfism), competition among managers, different priorities, unclear organizational objectives, failure to communicate strategic goals, inappropriate organizational structure, organizational culture, business unit culture.

6. Anyone directly involved in the project or in some way affected either positively or negatively as a result of the project. Stakeholders generally include the project manager, project team, project sponsor, and project customer and may include many others.

7. Schedule, Cost, Scope (can also be quality or performance specifications).

8. The triple constraint. Any change to one side of the triangle may affect the other sides.

9. Quality, availability of the project manager, timeliness of status reporting, reliability of the product or service deliverable, safety, minimum or mutually agreed upon scope changes, no impact or interruption to the work flow of the organization.

10. The project manager depends on the functional managers to provide the appropriate resources and to ensure that the work is performed correctly. A good relationship will minimize conflict and increase the likelihood of functional manager willingness to work on future projects with the project manager.

11. All project components and planning processes are interrelated. The project manager must coordinate and integrate project activities across organizational boundaries. The project manager ensures that functional units communicate effectively.

12. The functional manager provides the resources and technical expertise. Challenges include: different priorities among project managers and projects, managing the demands of multiple project managers, limited resources, unreasonable time frames, internal politics.

13. Establish expectations at the start of the project. Include communications requirements, escalation procedures, planning processes and methodology, and clear objectives.

14. There are many possible answers and solutions to the causes. This exercise is intended to emphasize the importance of identifying potential project problems and encourage proactive thinking and action.

15. The project manager is considered and integrator and coordinator for all major project activities. The project manager is held accountable for successful completion of the project. The project manager is a liaison between the project team and the project sponsor or executive steering committee. Other roles include – team builder, conflict manager, coach, mentor, facilitator, leader, motivator.

16. The project manager can develop better relationships with the functional managers through listening and understanding the priorities of the functional managers, their work environment, and issues associated with the functional manager's position.

17. Establish expectations clearly and intentionally between the project manager and the project sponsor or executive.

18. A project is a temporary endeavor undertaken to create a unique product, service, or result.

19. Initiating, Planning, Executing, Monitoring and Controlling, and Closing

20. Quality, resources, risk etc.

Your Personal Learning Library

Write down your thoughts, ideas, and observations about the material in the chapter that may assist you with your learning experience. Create action items and additional study plans to assist you in enhancing your skills or for preparing to take the PMP® or CAPM® exam.

Insights, key learning points, personal recommendations for additional study, areas for review, application to your work environment, items for further discussion with associates.

_____

_____

_____

_____

_____

_____

_____

_____

_____

_____

_____

Personal Action Items:	
Action Item	**Target Date for Completion**

Chapter Two

PROJECT MANAGEMENT GROWTH: CONCEPTS AND DEFINITIONS

This chapter focuses on the general evolution of project management from the 1940s through today's business environment. In the past four decades project management has grown from a means to achieve small project or activity completion to a key element in strategic planning. The use of powerful tools, enterprisewide methodologies, and processes to control activities and manage people and resources have become an accepted part of business management. The project life cycle, systems thinking, and the influence of project management processes on an organization's critical success factors have had a significant impact on overall organizational performance.

Glossary of terms   Key terms and definitions to review and remember

General Systems Management   A management technique designed to cross many organizational disciplines. For example: finance, manufacturing, engineering, and marketing.

Mature Project Management   The implementation of a standard methodology and accompanying processes that creates a high probability of repeated successes.

Product Scope   The features and functions that characterize the deliverable. This includes dimensions, features and physical characteristics.

Program   A group of related projects managed in a coordinated way to obtain benefits and control that is not available from managing them individually. Programs may include elements of related work outside the scope of the discrete projects in the program. Projects are normally the first-level subdivision of a program. Programs are also associated with ongoing activities.

Project Management Methodology   A repetitive process used on all projects to increase the likelihood of achieving project success, project management excellence, and maturity.

Project Scope   The work that must be accomplished to produce a deliverable with specified features and functions. The deliverable can be a product, service, or other result.

Stage-Gate Process   Stages are a group of activities that can be performed either in series or parallel based on the magnitude of risks the project team can endure. Gates are structured decision points at the end of each stage. These decision points are used to assess project performance and determine if corrective action is necessary.

System   A group of elements, either human or nonhuman, that is organized and arranged in such a way that the elements can act as a whole toward achieving some common goal or objective.

Activities, Questions, and Exercises

Refer to Chapter Two of *Project Management: A Systems Approach to Planning, Scheduling, and Controlling* (10th Edition) for supporting information and assistance in completing each exercise. The following questions and exercises are associated with the knowledge area of the *PMBOK® Guide*: The Project Management Framework.

Review each of the following questions or exercises and provide the answers in the space provided.

1. If you were assigned the task of developing a set of questions to assist an organization in determining if there is a need for a formal project management process, what questions would you ask?

_____

_____

_____

_____

2. As project management evolved it became apparent to many executives that a formal project management process did provide benefits to an organization. The driving forces of rapid technology changes, the increasing complexity of projects, and the increased demand for resources with specialized knowledge all contributed to the need for an effective project management methodology. How can a project management methodology assist in managing and resolving issues related to these driving forces? What benefits are obtained through the use of a project management methodology? What other driving forces may influence the decisions associated with establishing a project management methodology in an organization?

_____

_____

_____

3. Every organization encounters internal and external obstacles that may impede the achievement of success. For each obstacle listed, explain how a properly and effectively implemented project management methodology can overcome the obstacle.

Obstacle	Project management method, factor, or approach that may be used to respond to and overcome the obstacle
a) Resource shortages	

Obstacle	Project management method, factor, or approach that may be used to respond to and overcome the obstacle
b) Increasing costs	
c) Poor quality of work	
d) Technology changes	
e) Competition	

PM Knowledge Note

Project success depends on: setting objectives, establishing plans, organizing resources, staffing the team appropriately, establishing controls, and motivating the team. In addition, an understanding of the value expected to be achieved at project completion will impact the perception of project success.

4. Project management has been shown to significantly improve the overall performance of an organization. There are many approaches to managing projects, but providing the project manager with total integrative responsibility will result in some very specific advantages. Describe these advantages.

_____

_____

_____

_____

5. For each of the following items in the *Life Cycle Phases for Project Management Maturity* provide a recommendation for addressing and resolving the issue that will allow an organization to progress toward achieving project management maturity.

Project Management Maturity Life Cycle Phase	Issue	Recommendation
a) Embryonic	Recognizing the need	
b) Executive Management Acceptance	Executive understanding of project management	
c) Line Management Acceptance	Line management support	
d) Growth	Development of a project management methodology	
e) Maturity Phase	Developing an educational program to enhance project management skills	

6. Review the list of characteristics and match each characteristic with the appropriate industry classification.

Characteristic	Project Driven	Hybrid	Non–Project Driven
1. Project Manager has P&L responsibility	✓		
2. Primarily production driven but with many projects		✓	
3. Emphasis on new product development		✓	
4. Income comes from projects	✓		
5. Long life cycle products			✓
6. Large brick walls (functional structure)			✓
7. Multiple career paths	✓		
8. Marketing-oriented	✓	✓	
9. Short product life cycles		✓	
10. Project management is a recognized profession	✓		
11. Very few projects			✓
12. Need for rapid development process		✓	
13. Profitability from production			✓

7. To complete a project successfully a project manager is generally held accountable for a specific set of tasks. Place a check next to each item that is associated with a project manager's role during the project life cycle.

Set objectives   _____
Estimate activity costs   _____
Organize resources   _____
Estimate activity duration   _____
Establish plans   _____
Motivate personnel   _____
Provide staffing   _____
Define how tasks will be done _____
Issue directives   _____

Control each project task   _____
Remain flexible   _____
Establish project controls   _____

8. Few projects are completed without making decisions about trade-offs. What are the three elements that are generally considered when making trade-off decisions during project execution to achieve success?

   1. _____

   2. _____

   3. _____

9. Critical success factors (CSF) identify what is necessary to meet the customer's desired performance levels for project deliverables. Another form of success measurement is known as "Key Performance Indicators" (KPI). CSFs and KPIs are used to establish expectations and measurements of success for the work that must be performed to complete a project. Critical success factors and Key Performance Indicators should be established early in project planning and should be communicated to the key stakeholders at project start-up. In the table provided is a list of typical KPIs defined during project planning and monitored during implementation. In the spaces provided, identify the primary factors that are most commonly used to measure project success. What additional or secondary factors can be used as measurements of success?

Key Performance Indicators (quantifiable gauge that an organization uses to measure its performance in terms of meeting its critical success factors)
Use of a project management methodology
Establishment of control processes (cost, schedule, quality, etc.)
Use of interim or phase metrics—progress, trends, variances
Quality of resources assigned versus resources that were planned
Client involvement and feedback

Success Factors (Consider industry, strategic, and environmental factors)	
Primary	**Secondary**

10. **Quantitative and Qualitative failure.** Review each of the following items and identify which items are considered quantitative and which are considered qualitative failures:

Quantitative = 1
Qualitative = 2

Ineffective planning	1
Poor motivation	2
Poor productivity	1
Ineffective scheduling	1
Ineffective estimating	1
Poor morale	2
Project objectives changing or not defined clearly	1
Poor human relations	2
Lack of control processes	1
Lack of employee involvement and commitment	2
No functional management commitment	2
Conflicting priorities	2

11. Explain the stage-gate process and why it is necessary during project planning and execution.

_____

_____

12. What are the four common decisions associated with the gatekeeping process?

_____

_____

_____

13. Explain the difference between a product life cycle and a project life cycle.

_____

_____

_____

14. What is the generally accepted reason why projects should be managed using a project life cycle?

_____

_____

_____

15. Review each of the following characteristics of an effective project management methodology and explain why these characteristics are important to achieving project success.

Characteristic	Importance
Example: Standardized planning, scheduling, and cost control techniques	**Provides consistency in preparing project plans and increases the ability to monitor and control project performance.**
Use of templates	
Standardized reporting format	
Use of standardized life-cycle phases and end of phase reviews	
Flexibility of application to all projects	

16. It is important for project managers to be prepared to manage resistance to the application of a project management methodology. Explain why each of the following business units within an organization may resist the introduction of a project management methodology. Consider how you may respond to resistance in each case.

Business unit or entity	Reasons for resistance	Responses or actions that will minimize resistance
a) Sales		
b) Marketing		
c) Finance and Accounting		
d) Procurement		
e) Human Resources		
f) Manufacturing		
g) Engineering		

Kerzner "Quick tips" for the Project Management Institute PMP® and CAPM® EXAM

PROJECT MANAGEMENT

Why do we need project management?

- To determine roles and responsibilities
- Identify and establish priorities
- To manage deadlines and schedules
- To determine and assign accountability and authority
- Ensure proper documentation (determining what documentation is needed and to avoid overdocumentation)
- Provide a common methodology that will improve efficiencies in resource management
- Provide effective and timely status and communications
- Manage multiple projects more effectively
- Reduce rework and redundant work
- Capture and share lessons learned

Interpersonal skills   Interpersonal relationships are key elements to the success of any project. Project managers must utilize effective communication, the ability to influence project teams and functional groups to "get things done," leadership, motivation, conflict management, negotiation, and problem solving to manage teams and achieve project objectives.

The project environment   The project manager must maintain an awareness of the cultural and social environment, the international and political environment, and the physical environment when planning and implementing projects. These include such factors as education, ethnicity, economics, demographics, religion, time zone differences, regional holidays, and the physical surrounding that may affect how the project is managed. These elements are associated with the domain of project management known as "Professional Responsibility."

Progressive elaboration   Proceeding in steps and adding more detail.

Phases and project life cycles   We divide projects into phases to provide better management control. Project life cycles generally define what technical work will be done in each phase, when a deliverable is required, who will be involved in each phase, and how each phase will be controlled. The project life cycle may also be used to define how a phase and the associated deliverables will be approved.

Cost and staffing   These are generally low at the start of a project, increase through each progressive phase, peak during the intermediate phases, and then drop rapidly as the project reaches closure.

Risk and uncertainty   Generally risk and uncertainty are considered to be at the greatest level during the start of the project but the certainty of project success progressively improves through each phase (if proper planning continues and the appropriate and necessary control processes are in place).

Project-based organizations   Derive their revenue from performing projects for other organizations through contractual agreements or organizations who have adopted a management by project approach using systems that have been established for tracking and reporting on multiple, simultaneous projects.

Non–project-based organizations   Do not have the management systems in place to manage and support projects. These organizations may have subunits that operate as project-based organizations but the absence of project-oriented systems at the enterprise level and the type of organizational structure in place may make project management more difficult.

Answers to Questions and Exercises

1. How often are projects completed on time, within budget, and according to performance specification? How effective is the communication between functional units? What factors contribute to project failure? How can project management be used to improve overall performance? What is the level of customer satisfaction regarding completed projects? How are projects selected? How can resource management be improved?

2. Prioritization of projects. More effective management of changes. More effective utilization of available resources. Improved coordination between functional units.

3. a) Prioritization of projects, resource planning, effective schedule development.
   b) Change control process, cost estimating techniques, risk management.
   c) Establish a quality policy, schedule reviews and inspections, plan for requirements management.
   d) Phase end reviews, scope management, change control and configuration management.
   e) Quality control, cost management, effective planning, trade-off analysis.

4. Total accountability is assumed by a single person. Project rather than functional dedication. Coordination across functional interfaces can be achieved. Improved utilization of integrated planning and control processes.

5. a) Explain the benefits of repeatable processes, identify areas for improvement in schedule, cost and requirements management.
   b) Provide supporting information about project successes and the use of best practices.
   c) Obtain and communicate management support to functional managers, explain the benefits of project management, acknowledge the needs of the functional managers.
   d) Establish a PMO, standardize methods, establish enterprisewide processes, provide project management training.
   e) Establish project management as a career path, continuous improvement through reviews and feedback, provide training to all levels of employees, maintain management support and visibility, document lessons learned.

6. Project driven—1,4,7,8,10
   Hybrid—2,3,8,9,12
   Non-project driven—5,6,11,13

7. Set objectives, organize resources, establish plans, motivate personnel, provide staffing, issue directives, remain flexible, establish project controls.

8. Cost, Time, Scope.

9. Primary—Within schedule, within cost, within quality requirements, accepted by the customer. Secondary—Follow-on work, customer authorizes use of the name for references, minimum or mutually agreed upon scope changes, no disturbance of normal operations, within safety requirements, managed fairly and ethically, supported corporate strategies, maintained corporate reputation or brand name, within government regulations.

10.

Ineffective planning	1
Poor motivation	2
Poor productivity	1
Ineffective scheduling	1
Ineffective estimating	1
Poor morale	2
Project objectives changing or not defined clearly	1
Poor human relations	2
Lack of control processes	1
Lack of employee involvement and commitment	2
No functional management commitment	2
Conflicting priorities	2

11. The stage gate process establishes reviews at the completion of each phase to determine if the project should continue on to the next phase. Also used to determine where corrective action may be necessary to return project performance to acceptable levels.

12. Proceed to the next phase based on objectives, revise objectives and then proceed to next phase, correct variances before moving to the next phase, cancel the project based on current status and other factors.

13. The project life cycle includes the work required to produce the product or the project. The project life cycle ends when the product has been delivered. The product life cycle includes the R&D, project life cycle, operations and maintenance, and eventual termination or salvage of the product.

14. More effective manageability and control.

15. Use of templates—more efficient use of time, consistency, Standardized reporting format—consistency, facilitates analysis, Use of standardized life-cycle phases—consistency in planning and control, Flexibility of application to all projects—all projects are unique. Some processes may not apply to all projects.

16. a) Potential loss of power and position with the customer, credit given to the project manager.
   b) Potential for the project managers to become associated with the marketing function. Customer has more contact with the project manager. Fear of loss of credit for sales.
   c) Development of independent project accounting systems, loss of control.
   d) Bypassing of the established processes.
   e) Potential additional workload, new job descriptions, more training requirements.
   f) Potential loss of experienced resources to the project.
   g) May not agree with the methodology, may not support the planning process, may perceive project management as overhead.

Your Personal Learning Library

Write down your thoughts, ideas, and observations about the material in the chapter that may assist you with your learning experience. Create action items and additional study plans to assist you in enhancing your skills or for preparing to take the PMP® or CAPM® exam.

Insights, key learning points, personal recommendations for additional study, areas for review, application to your work environment, items for further discussion with associates.

_____

_____

_____

_____

_____

_____

_____

_____

_____

_____

_____

Personal Action Items:	
Action Item	**Target Date for Completion**

Chapter Three

ORGANIZATIONAL STRUCTURES

The organizational structure, how departments, business units, resources, responsibilities, and job assignments are arranged, is an important factor in the project planning process and may significantly impact project communication, the authority of the project manager, and the process of obtaining project human resources.

Organizational structures vary by company and create unique challenges for the project manager. The effectiveness of the project manager and the ability to coordinate activities cross-organizationally is directly associated with how an organization is structured. Type of functional expertise, management hierarchy, authority levels, and conflicts associated with differing priorities and cultures within an organization require special handling and, in many cases, strong interpersonal and negotiating skills.

Glossary of terms Key terms and definitions to review and remember

Accountability   To be answerable for the satisfactory completion of a specific assignment.

Authority   The power granted to individuals (possibly by their positions) that enables them to make final decisions.

Customer/User   The person or organization that will use the final product or service of the project.

Organizations   Groups of people and their associated resources who must coordinate their activities in order to meet specific objectives.

Organizational Structures   There are three main types of organizational structure.

> **Functional**   Also known as chimney, smoke stack, or silo structure. In this structure there is generally one manager overseeing the work and the type of work is associated with a specific skill or expertise. Examples—accounting, marketing, engineering, manufacturing.

> **Matrix**   This type of structure is designed to maximize the use of resources by assigning resources to multiple functions. This creates an environment where employees may report to two or more managers. In this type of structure there will be horizontal as well as vertical channels of communication and negotiation for resources expected between project managers and functional groups.

Pure Project or Projectized   The organization is arranged by project with each project manager assigned total authority and accountability for the project. All project staff and personnel report directly to the project manager.

Performing Organization   The business unit, enterprise, or entity whose employees are most directly involved in doing the work of the project.

PMO   Project Management Office—an organizational unit established to centralize and coordinate the management of projects under its domain.

Portfolio   A collection of projects or programs and other work that are grouped together to facilitate effective management of that work and those projects and programs to achieve strategic objectives.

Project Coordinator   This position is generally assigned by a higher level manager such as a division manager, business unit vice president, or the CEO, and reports to that manager. The position is granted some authority for decision making.

Project Expeditor   Generally found in a functional organizational structure. The expeditor is a type of staff assistant with no actual authority for decision making. This position attempts to communicate information across functional boundaries.

Responsibility   The obligation incurred by individuals in their roles in the formal organization to effectively perform assignments.

Activities, Questions, and Exercises

Refer to Chapter Three of *Project Management: A Systems Approach to Planning, Scheduling, and Controlling* (10th Edition) for supporting information. Review each of the following questions or exercises and provide the answers in the space provided.

The following questions and exercises are related to the *PMBOK® Guide* Knowledge Area—Project Human Resources Management and the Project Framework

Review each of the following questions or exercises and provide the answers in the space provided.

1. For many years most organizations utilized a hierarchal structure for managing operations and personnel. This traditional structure was based on divisions, departments, and functional units.

   Describe the advantages and disadvantages of the traditional or classic organizational structure: (The traditional structure is generally associated with several levels of management in a specific hierarchy with many levels between the lowest position and highest position.)

Advantages	Disadvantages

2. Which of the following organizational structures would provide the most effective platform for building a high performance team? Provide the rationale to support your answer.

Functional _____ Matrix _____ Projectized or Pure Project _____

Rationale: _____

_____

_____

3. ✍ **PM Quick Check:** Functional, Matrix, and Projectized (or Pure Project) Organizational Structures.

Match the type of structure to the correct characteristics or description.

Functional	Weak Matrix	Balanced Matrix	Strong Matrix	Projectized or Pure Project

1. Project manager has little to no authority

2. Project expeditor

3. Project manager has some authority

4. Project manager has high to total authority

5. Equal authority between project manager and functional manager

6. Project manager has a moderate to high level of authority

7. Project coordinator

8. One clearly defined manager

9. Staff members grouped by specialty

10. Project work is done independently

11. Team members are often co-located

12. Resources report directly to the project manager

13. No control of the project budget

14. Full control of the project budget

PM Knowledge Note	*Project success in a matrix structure is more likely to be achieved if there is strong coordination and effective communication between the functional groups interacting cross-organizationally. This will help to ensure that project goals and objectives are understood at the functional and project level. The project manager creates the integrative environment and cross – organizational relationships that will drive project success.*

4. Matrix Organizational Structure. Consider the Matrix organizational structure. Describe the primary advantages and potential disadvantages if this type of structure is selected by an organization.

Advantages	Disadvantages

5. Explain the difference between authority, responsibility, and accountability:

Authority	
Responsibility	
Accountability	

6. Define the term "stakeholder" and provide examples of stakeholders that are typically associated with projects.

_____

_____

_____

7. Why is organizational culture considered to be an "enterprise environmental factor?"

_____

_____

8. Complete the following table by describing the influences and relationships between the project characteristics and the organizational structure: Refer to the *PMBOK® Guide*—Fourth edition – Organizational Structure

Project Characteristic	Functional Structure	Weak Matrix	Balanced Matrix	Strong Matrix	Projectized
Project manager authority	Little or none				
Resource availability					
Who controls the project budget					Project Manager
Project manager role					
Project Management Administrative support					

The Project Management Center of Excellence—A Review

Some organizations have created support groups to develop methodologies, templates, training programs, standards, and review processes to ensure consistency in the management of approved projects. These centers of excellence are also referred to as PMOs or Project Management Offices. PMOs may be created to coordinate activities associated with large complex projects or to manage several projects in a portfolio. A PMO may also be referred to as a "project office" or "program office."

The PMO can support an organization by:

- Centralized coordination and communication across multiple projects
- Developing enterprisewide project management policies, procedures, templates, and documenting lessons learned
- Prioritization of projects and effective management of critical resources
- Centralized monitoring and standardized performance reporting for all projects
- Consolidating project progress and performance reporting for executive management
- Providing training, mentoring, and coaching for project managers
- Developing a historical database of lessons learned from exiting and previous projects
- Analyzing overall project risks, opportunities, and the interactions between projects

Kerzner "Quick tips" for the Project Management Institute PMP® EXAM

Remember the three types of organizational structure—functional, matrix, and projectized (also known as pure project).

There are three types of matrix structures—Weak, Balanced, and Strong. The project manager has the greatest level of authority in the strong matrix. In the balanced matrix, the project manager and functional managers share an equal level of authority and there is a potential for conflict due to a "two boss" reporting structure for the resources assigned to the project.

Accountability   The project manager commits to achieve the project objectives and is answerable for overall project performance.

Authority   Generally associated with position or rank and granted or assigned by senior or executive management.

Study note:

Basic factors that influence the selection of a project organizational form:

1. Project size and complexity

2. Project length (duration)

3. Experience with project management organization and methodologies

4. Philosophy and visibility of upper-level management

5. Project location

6. Available resources

7. The unique aspects about the project (the deliverables, new technologies, visibility, type of resources to be used, experience of the project manager, stakeholder needs and expectations, impact of the project on organizational objectives)

Answers to Questions and Exercises

1. Advantages—Specific manager, budgeting and cost control are easier to manage, effective control of technical issues and expertise, flexibility in the use of people resources, communications channels are well established, control over personnel, continuity of policies and procedures.

   Disadvantages—No specific responsibility for a project, coordination is difficult, decisions are based on politics and the influence of the strongest functional group, response to customer issues may be slow, tendency toward functional loyalty instead of project loyalty, communication through levels is slow and may become distorted.

2. Projectized or Pure Project. The project manager has the greatest level of authority and influence over the team. The team is a dedicated resource to the project with no other priorities. The project team performance reviews are linked to project success.

3. Functional—1,2,8,9,13, Weak Matrix—3,7, Balanced Matrix—5, Strong Matrix—6, Projectized—4,10,11,12,14

4. Advantages—Better utilization of resources, improved coordination between functional units, expertise of functional units is used more efficiently, project manager acts as a coordinator and may have some authority.

   Disadvantages—Functional resources must manage multiple priorities and demands, resources may become overallocated, schedule conflicts, conflicts between project manager and functional manager about task assignments and control of work.

5. Authority – the formal, justified or legitimate right granted to an individual (generally by position or rank) to make decisions and exercise available power to achieve a desired end. Responsibility – The obligation to effectively perform assigned tasks. Accountability - the acknowledgment and assumption of responsibility for actions, products, decisions, and policies including the administration, governance and implementation within the scope of the role or employment position and encompassing the obligation to report, explain and be answerable for resulting consequences.

6. A stakeholder is defined as any person or organization directly involved in or in some way affected either positively or negatively as a result of a project. Examples of stakeholders – project manager, sponsor, customer, contractor, supplier, business unit.

7. The culture of an organization and the cultures of specific are associated with norms that have been developed over time and may influence the project planning process. Refer to the *PMBOK® Guide—Fourth edition* – Organizational Cultures and Styles

8.

Project Characteristic	Functional Structure	Weak Matrix	Balanced Matrix	Strong Matrix	Projectized
Project manager authority	Little or none	Limited	Low to moderate	Moderate to high	High to almost total
Resource availability	Little to none	Limited	Low to moderate	Moderate to high	High to almost total
Who controls the project budget	Functional manager	Functional Manager	Mixed	Project manager	Project Manager
Project manager role	Part time	Part time	Full time	Full time	Full time
Project Management Administrative support	Part time	Part time	Part time	Full time	Full time

Your Personal Learning Library

Write down your thoughts, ideas, and observations about the material in the chapter that may assist you with your learning experience. Create action items and additional study plans to assist you in enhancing your skills or for preparing to take the PMP® or CAPM® exam.

Insights, key learning points, personal recommendations for additional study, areas for review, application to your work environment, items for further discussion with associates.

_____

_____

_____

_____

_____

_____

_____

_____

_____

_____

Personal Action Items:	
Action Item	**Target Date for Completion**

Chapter Four

ORGANIZING AND STAFFING THE PROJECT OFFICE AND TEAM

Achieving the objectives of a project and ensuring customer satisfaction requires a project team that has the necessary skills, is fully committed to the project, and is managed by a project manager who possesses effective organization, communication, and leadership skills. Staffing the project with the appropriate team members is often a challenge due to organizational policies, organizational structure, and availability of resources. The project manager generally attempts to obtain the most highly qualified team members but must often accept competent performers who must receive additional training and may require additional supervision, coaching, and mentoring. Organizing the project team after selection by establishing a project organization chart will facilitate the communications between team members and clearly identify roles, responsibilities, and linkages between project team members.

Glossary of terms   Key terms and definitions to review and remember

Competency   The skill or ability and capacity required to complete assigned project activities.

Organization   A group of people brought together for a specific purpose or to perform some type of work within an enterprise.

Organization Chart   A method for depicting interrelationships among a group of people working together toward a common objective.

PMO   Project Management Office—an organizational unit established to centralize and coordinate the management of projects under its domain.

Project Charter   A document issued by the project initiator or sponsor that formally authorizes the existence of a project and provides the project manager with the authority to apply organizational resources to project activities.

Project Organization Chart   A graphic display of project team members and their reporting relationships. Detail depends on project complexity and size of the project team.

Project Role   This is the label or specific function assigned to a project team member or stakeholder. It describes the portion of a project that will be completed, or the function that will be performed by the individual.

Project Team   The personnel assigned to the project including the project manager, functional managers, and in some cases the project sponsor and customer.

Activities, Questions, and Exercises

Refer to Chapter Four of *Project Management: A Systems Approach to Planning, Scheduling, and Controlling* (10th Edition) for supporting information. Review each of the following questions or exercises and provide the answers in the space provided.

The following questions and exercises are related to the *PMBOK® Guide* Knowledge Area —Project Human Resources Management and the Project Framework

1. Preparing to Staff a Project Team or Project Office

   A. List at least eight personal characteristics or skills a project manager should possess to be successful in managing a project team and achieving project objectives.

      1.

      2.

      3.

      4.

      5.

      6.

      7.

      8.

   B. The project manager is usually selected by the

      _____

   C. What characteristics and personality traits would be desirable to you when selecting your project team? List at least six traits or characteristics.

      1.

      2.

      3.

      4.

5.

6.

D. What problems or obstacles may be encountered by the project manager when attempting to staff his or her project team? List at least three problems or issues that may be experienced when staffing a project and provide a recommendation for managing and resolving the issue. Consider the project you are working on. What issues have you encountered and how did you respond to those issues?

Issue or Problem	Recommended Action
Example: Critical resources may not be available when needed	Ensure effective scheduling of resources. Revise the project network diagram if possible to accommodate resource availability. Identify back-up sources, escalate to the project sponsor, negotiate availability with functional managers

2. ✎ **PM Quick Check:** In which organizational structure will the project team members encounter a situation known as the "two boss syndrome?"

_____ Functional

_____ Matrix

_____ Projectized

PMP® & CAPM® Exam

3. Describe the major responsibilities of the project manager:

1. _____

2. _____

3. _____

4. _____

5. _____

6. _____

7. _____

8. _____

9. _____

4. In addition to the desired personal characteristics, a project manager must fully understand the needs of the project team members and provide the appropriate level of support to ensure the team has the ability to perform the project work efficiently. What information and expectations should the project manager communicate to the project team?

1. _____

2. _____

3. _____

4. _____

5. _____

6. _____

7. _____

8. _____

5. A team will perform more effectively if the project manager provides a positive atmosphere that is conducive to teamwork. Describe the actions a project manager may take to ensure he or she creates a high-performing team.

_____

_____

_____

_____

_____

_____

_____

_____

_____

6. Selecting the wrong person as project manager. The following criteria are sometimes used to select and assign a project manager. For each item listed, explain why it may result in the assignment of the "wrong project manager."

Criteria	Explanation
Maturity (age)	
Hard-nosed tactics (forcing, controlling)	
Availability (waiting for an assignment)	
Technical expertise (technically proficient)	

Criteria	Explanation
Customer orientation (customer desires a person because they are familiar with him/her)	
New exposure (to become familiar with the project environment)	
Company exposure (the person has experience in many areas or business units with the organization)	

7. ✎ **PM Quick Check:** What is the definition of a "stakeholder?"

A stakeholder is

PMP® & CAPM® Exam

8. **Managing difficult people.** Many project managers experience situations in which some team members become difficult to manage or create conflicts that may disrupt the entire project team and jeopardize the project. These situations include but are not limited to:

Nonacceptance of rules, policies, and procedures
Nonacceptance of established formal authority
Focusing on technical aspects at the expense of the budget and schedule
Failing to contribute to the planning process
Delays in providing project information
Incompetence or inability to perform assigned activities

What actions could be taken by the project manager to resolve these issues? There may be many answers to this question.

_____

_____

_____

_____

9. Why is it important for the project manager to prepare and distribute a project organization chart to the project team and other stakeholders? Provide at least three reasons.

_____

_____

_____

_____

10. ✍ **PM Quick Check:** What is the project manager's primary function?

> Project manager's primary function is

PMP® & CAPM® Exam

PM Knowledge Note

The staffing process usually involves asking these questions:
What people resources are required? What skill levels are needed?
Where will the people come from? When will the resources be needed?
What type of project organizational structure will be best?

---

11. **Selecting the project team.** A challenge faced by most project managers is the selection of the "right" people for the project team and the use of appropriate managerial, motivational, and leadership skills to keep the team focused on project objectives and the completion of deliverables. This means managing a variety of personality types. Review each type of person that may become part of the project team and match the characteristics of each type.

Personality Type	Matching Letter	Characteristic or phrase often used	Desired team member Y or N
1. Aggressor		A. Cannot focus on ideas for a long time unless it's his/her idea	
2. Dominator		B. Finds fault in all areas of the project/ project management	
3. Initiator		C. Likes to hear himself/herself talk. Likes to boast	
4. Topic jumper		D. Criticizes everyone. Deflates the status and ego of other team members	
5. Information seeker		E. There is a chance this may work. Let's try this!	
6. Encourager		F. Rejects the views of others. Let me try to put this in perspective. Are we saying that….?	
7. Withdrawer		G. Your idea has merit. What you said will really help!	

Personality Type	Matching Letter	Characteristic or phrase often used	Desired team member Y or N
8. Clarifier		H. Obtains overall team agreement.	
9. Recognition seeker		I. Let me try to put this in perspective. Are we saying that….?	
10. Information giver		J. Is afraid to be criticized. May be shy. Does not participate openly.	
11. Blocker		K. Always tries to take over. Manipulates people. Challenges the leader or person in charge.	
12. Harmonizer		L. The literature says…… Here is the documentation. Benchmarking shows that….	
13. Consensus taker		M. How can we obtain lessons learned or best practices? Where can we find what we need?	
14. Devil's advocate		N. Your ideas and mine are very similar. How can we agree on this?	

12. According to the *PMBOK® Guide*—Fourth edition, acquiring the project team is associated with four basic tools and techniques: pre-assignment, negotiation, acquisition, and the use of virtual teams. What are the potential difficulties associated with each of these techniques and what actions may be taken by the project manager to minimize these difficulties? There are many possible responses for this exercise.

Tool / technique	Difficulties/Challenges	Possible Solutions / Responses
Pre-assignment	Example: No input from the project manager during the staff selection process.	Develop a working relationship with the team using effective communications and interpersonal skills
Negotiation		

Acquisition		
Virtual teams		

13. Provide examples of actions and activities a project manager may take to ensure an environment and project atmosphere that will encourage high levels of teamwork and a cohesive team.

_____

_____

_____

_____

_____

_____

14. According to Dr. Harold Kerzner, what are the 10 major skill requirements for project and program managers?

a.

b.

c.

d.

e.

f.

g.

h.

i.

j.

15. Managing conflicts:

Match each type of conflict management technique with the appropriate definition

Withdrawing /avoiding _____	A. Offers only win / lose solutions
Smoothing /Accommodating _____	B. Emphasize areas of agreement instead of areas of disagreement
Compromising _____	C. Give and take attitude and dialog
Forcing _____	D. A win lose / lose win situation in which some satisfactions is gained by both sides
Collaborating _____	E Retreating from a potential conflict
Confronting/ Problem Solving _____	F. Incorporating multiple viewpoints and achieving consensus

**Kerzner "Quick tips" for the Project Management Institute PMP®
and CAPM® EXAM**

HUMAN RESOURCES MANAGEMENT

The four processes associated with Project Human Resources management are:

> Develop Human Resource Plan
>
> Acquire Project Team
>
> Develop Project Team
>
> Manage Project Team

Human resource planning   These processes determine project roles, responsibilities, and reporting relationships in the project environment.

Staffing management plan   This is a subset of the project management plan and describes when and how human resource requirements will be met. The staffing management plan is developed based on the specific needs of the project. The plan includes processes for acquisition of team members, a timetable for obtaining or recruiting team members, the allocation of work among the resources, and the release criteria. Training requirements, recognition and reward procedures, safety requirements, and compliance requirements are also generally included in the staffing management plan. The plan also addresses the impact of the project staffing process on the organization.

Organizational theory   This is associated with how different organizational structures affect the behavior of an organization. Remember the functional, matrix, and projectized structures and how these structures may affect the project environment.

Acquiring the project team   Consider the enterprise environmental factors such as availability, experience, and cost. There are organizational processes to follow for recruiting or hiring and roles and responsibilities should be clearly defined to minimize miscommunication and conflict.

Developing the team   After the team has been formed, the project manager must focus some energy and attention on building the team. A well-managed and motivated team is more likely to commit to the project objectives than a loosely formed group. Techniques for team building include: general management skills as well as interpersonal or soft skills, training, team-building activities such as off site meetings and events, team ground rules established early in the project, co-location (when applicable and feasible), recognition and reward to maintain the motivation.

Additional tips and practice items for the PMP® exam are included in each chapter and in the section of the workbook entitled ***PMP Exam and PMBOK Guide® Review***.

Answers to Questions and Exercises

1. A)

    1. Honesty and integrity

    2. Understanding of personnel problems

    3. Understanding of project technology

    4. Business management competence

    5. Alertness and quickness

    6. Versatility

    7. Energy and stamina (toughness)

    8. Decision-making ability

    9. Ability to evaluate risk and uncertainty

    10. Communications

    11. Planning

       12. Interface management

       13. Team building

       14. Leadership

       15. Conflict resolution

       16. Organizing

       17. Creativity and entrepreneurship

B) The project manager is usually selected by *the project sponsor* or *project executive.*

C) Technical ability and competence, willingness to work with the team, willingness to accept changes, flexibility, good listener, honesty, reliability, understanding and acceptance of their responsibilities, accurate.

D) Answers provided in the following table.

Issue or Problem	Recommended action
Example: Critical resources may not be available when needed	Effective scheduling of resources. Identify backup sources, escalate to the project sponsor, negotiate availability with functional managers
Multiple projects result in resource contention	Prioritize projects to ensure that highest priority projects receive the appropriate resources
Functional managers unwilling to release highly qualified resources	Explain relationship of the project to organizational objectives. Negotiate time frames to minimize length of time the resources are needed
Lack of resources with subject matter expertise	Subcontract the work, provide training, hire additional resources
Resources reluctant to accept project assignment due to the temporary nature of projects (start and end dates)	Bonuses and rewards at project completion, job placement support, training for other positions, negotiate return of the resource with functional managers
Performance appraisals remain the responsibility of the functional manager	Negotiate joint performance appraisal input and supervisory responsibility

2. Matrix structure.

3. Integration of project work across functional lines, ensure the project plan is developed, direct the preparation of the project budget, direct the preparation of the project schedule, establish monitoring and control procedures, create the project organization chart, explain duties and responsibilities of the project team, set expectations with the project team and sponsor, select the project team, communicate project objectives, estimate resource requirements, direct the project work, develop a high-performing team.

4. Project objectives, contractual requirements, quality requirements, constraints, assumptions, customer/client information, relationship of the project to organizational objectives, status reporting, change management process, milestone dates, sponsor expectations.

5. Clearly defined roles and responsibilities, empower the team members to perform their responsibilities, regularly reward and recognize excellent job performance, providing coaching and mentoring, avoid aggressive control of the team members, avoid doing functional manager work, establish a supportive environment, avoid personal recognition and assuming credit for the work done by the team, ensure the team receives credit for achievements, invite participatory decision making when possible.

6.

Criteria	Explanation
Maturity (age)	Age is not always an indicator of strong experience. Age is often confused with experience. Decisions should be made based on performance and competency.
Hard-nosed tactics (forcing, controlling)	This type of manager may achieve acceptably in the short term but may create interpersonal conflicts that will affect project performance later including low morale and poor productivity.
Availability (waiting for an assignment)	An "available" project manager may not have the appropriate skill set or experience. Sometimes known as the accidental project manager.
Technical expertise (technically proficient)	Technically strong project managers may tend to manage the technical details of the project instead of the cross-organizational issues and overall integration of project deliverables.
Customer orientation (customer desires a person because they are familiar with him/her)	The person selected may have a personality desired by the customer but may not possess the necessary project management skills for the project.
New exposure (to become familiar with the project environment)	Some projects, due to their complexity, may not be suitable for use as "training ground" for new project managers. This type of assignment may be perceived as short term and may result in a lack of commitment from the project team.
Company exposure (the person has experience in many company areas and business units within the organization)	This type of action is often referred to as "ticket punching." The assignment is associated with career advancement and the project manager may not be fully committed to the project. This type of assignment may be perceived as short term and may result in a lack of commitment from the project team.

7. PM Quick Check. A stakeholder is any person or organization that is directly involved in, or some way impacted either positively or negatively, as a result of the project.

8. Screen project personnel before adding them to the team. Set expectations early, establish roles and responsibilities early, explain the integrative project management process, communicate the importance of teamwork, set vision and goals.

9. The organization chart defines the various relationships within the project team and defines at a high level the roles and responsibilities of the team members. It assists in the overall communications process with the project team and organizes the project.

10. **PM Quick Check.** Project manager's primary function is the integration of project activities.

11. Selecting the project team.

Personality Type	Matching Letter	Characteristic or phrase often used	Desired team member Y or N
1. Aggressor	D	A. Cannot focus on ideas for a long time unless it's his/her idea. Must be first with new ideas	N
2. Dominator	K	B. Finds fault in all areas of the project/project management	N
3. Initiator	E	C. Likes to hear himself/herself talk. Likes to boast	Y
4. Topic jumper	A	D. Criticizes everyone, attacker. Deflates the status and ego of other team members	N
5. Information seeker	M	E. There is a chance this may work. Let's try this!	Y
6. Encourager	G	F. Rejects the views of others. Let me try to put this in perspective. Are we saying that….?	Y
7. Withdrawer	J	G. Your idea has merit. What you said will really help!	N
8. Clarifier	I	H. Obtains overall team agreement.	Y
9. Recognition seeker	C	I. Let me try to put this in perspective. Are we saying that….?	N
10. Information giver	L	J. Is afraid to be criticized. May be shy. Does not participate openly	Y
11. Blocker	F	K. Always tries to take over. Tries to manipulate people. Challenges the leader or person in charge	N
12. Harmonizer	H	L. The literature says…… Here is the documentation. Benchmarking shows that….	Y

Personality Type	Matching Letter	Characteristic or phrase often used	Desired team member Y or N
13. Consensus taker	N	**M.** How can we obtain lessons learned or best practices? Where can we find what we need?	Y
14. Devil's advocate	B	**N.** Your ideas and mine are very similar. How can we agree on this?	N

12.

Tool / technique	Difficulties/Challenges	Possible Solutions / Responses
Pre-assignment	No input from the project manager during the staff selection process. Lack of skilled resources, Project manager personality may conflict with the established team	Establish clearly defined roles and responsibilities as early as possible. Develop a working relationship with the team using effective communications and interpersonal skills. Obtain information about each project staff member and schedule individual meetings to establish expectations. Meet with the entire team and develop joint expectations
Negotiation	Functional managers unwilling to negotiate. Competition for the same resources between multiple project managers	Obtain training in effective negotiation skills. Listen to the needs of the functional managers and develop an understanding of their view of priorities

Acquisition	Unreliable resources, lack of skills among acquired staff, cost,	Identify qualified contractors and resources. Define specific skill requirements. Identify multiple suppliers of resources. Establish clearly defined quality standards and expectations
Virtual teams	Difficulty in maintaining communication, time zones, technology issues	I Define technology requirements, schedule frequent virtual meetings,. Assess the diversity of the team and establish a communications plan that will meet the needs of the virtual team.

13. Teambuilding activities, co-location when possible, training (as a team when feasible),, recognition and reward, develop a team charter, demonstrate honesty and integrity, establish a workable and effective communications plan, set expectations,

14. a) teambuilding   b) Leadership   c) Conflict resolution   d) Technical expertise
    e) Planning skills   f) Organizational skills   g) Entrepreneurship   h) administrative
    skills   i) Ability to gain management support   j) Resource allocation skills

15. Managing conflicts: Match each type of conflict management technique with the appropriate definition

Withdrawing / avoiding	E	A. Offers only win / lose solutions
Smoothing / Accommodating	B	B. Emphasize areas of agreement instead of areas of disagreement
Compromising	D	C. Give and take attitude and dialog
Forcing	A	D. A win lose / lose win situation in which some satisfactions is gained by both sides
Collaborating	F	E Retreating from a potential conflict
Confronting / Problem Solving	C	F. Incorporating multiple viewpoints and achieving consensus

Your Personal Learning Library

Write down your thoughts, ideas, and observations about the material in the chapter that may assist you with your learning experience. Create action items and additional study plans to assist you in enhancing your skills or for preparing to take the PMP® or CAPM® exam.

Insights, key learning points, personal recommendations for additional study, areas for review, application to your work environment, items for further discussion with associates.

_____

_____

_____

_____

_____

_____

_____

_____

_____

_____

Personal Action Items:	
Action Item	**Target Date for Completion**

Chapter Five

MANAGEMENT FUNCTIONS

General management functions are closely related to the functions of the project managers and, in many cases, overlap with the roles and responsibilities of the project manager. It is important for the project manager to fully understand the connection between his or her role as the key integrator for planning and effective management of the project team. To be successful, project managers must be aware of the total project environment. This includes the cultural and social environment, the international and political environment and the physical environment. This understanding combined with general management skills, interpersonal skills, and the appropriate application of project management skills, tools and techniques will establish a foundation for planning, team building, problem resolution, and an increased probability of successful project management depends on an understanding of the team's physiological needs and the behavior displayed when people perform in groups and in teams.

PM Knowledge Note

*An understanding of the social, cultural, and international factors that may affect a project is necessary in today's global project environment and is directly related to **professional and social responsibility**.*

Glossary of terms   Key terms and definitions to review and remember

Accountability   An obligation or willingness to accept the consequences and results of project success or failure.

Authority   The legal or rightful power to command, act, or direct the activities of others. The right of an individual to make necessary decisions required to achieve objectives.

Communications

> **Downward**   Usually provides direction to and control of the project team. It is generally associated with job-related information such as activities, schedules, evaluation, and feedback.

Lateral   Horizontal flow of information to peers, functional groups, contractors, clients, and support personnel. Most of the communications activities associated with the project manager position are lateral.

Upward   Primarily associated with information of interest to high-level management for assessing overall project performance. Exception reporting is sometimes used to provide project status or feedback about production or customer satisfaction issues.

Controlling   A process for measuring progress toward an objective, evaluating what remains to be done, and taking the necessary corrective action to achieve objectives.

Directing   The implementation and execution of approved plans through others to achieve or exceed objectives.

Empirical   Data obtained by studying the experiences of others.

Leadership   Developing a vision and strategy and motivating people to achieve that vision. The ability to influence and cause people to follow.

Autocratic Leadership   Strong focus on task completion with very little concern for the project team. High level of formal authority and makes or approves all decisions.

Laissez-Faire Leadership   The leader or project manager assigns all work to the project team or workers. The project manager makes occasional appearances but remains mostly inactive and not involved in day-to-day management.

RAM   Responsibility Assignment Matrix. The RAM is used to align major project tasks with the individuals associated with each task. The RAM assists the project manager and team in defining clear responsibility for project tasks.

Responsibility   The assignment of a specific event or activity and the expectation to complete that task.

Activities, Questions, and Exercises

Refer to Chapter Five of *Project Management: A Systems Approach to Planning, Scheduling, and Controlling* (10th Edition) for supporting information. Review each of the following questions or exercises and provide the answers in the space provided.

The following questions and exercises are associated with the knowledge areas of the *PMBOK® Guide*: Project Human Resources Management and Project Communications Management. Some elements of Professional and Social Responsibility are also addressed in this chapter.

1. List five principal functions of management.

   a. _____

   b. _____

   c. _____

   d. _____

   e. _____

2. Match the term with the appropriate explanation:

Term	Explanation
Measuring	
Evaluating	
Correcting	

... resolve an unfavorable trend or deviation or taking advantage of an unusually favorable trend

... determining through formal or informal reports the degree to which progress toward objectives is ...

... determining the best possible responses to significant deviations from planned performance.

3. Complete ... project manager is responsible for ensuring the

General management encompasses planning, organizing, staffing, and controlling the operations of an ongoing enterprise.

4. Exercise: *From the director's chair*

- As a project manager you will occasionally find yourself in the role of director (someone who supervises the actors and directs the action in the production of a show). Considering your role as director review each of the following terms and match them with the appropriate explanation.

Staffing	Training	Supervising	Delegating

Motivating	Counseling	Coordinating

A. Giving day-to-day instruction, guidance, and discipline
B. Seeing that activities are carried out in relation to their importance and with minimum conflict

(*continued on next page*)

C. Holding private discussions with another about how he or she might do better work, solve a personal problem, or realize his or her ambitions

D. Seeing that a qualified person is selected for each position

E. Encouraging others to perform by fulfilling or appealing to their needs

F. Assigning work, responsibility, and authority so others can make maximum utilization of their abilities

G. Teaching individuals and groups how to fulfill their duties and responsibilities

---

Theory X, Theory Y, and Theory Z
Understanding the effect of managerial styles on a project team is extremely important. The style of management and the motivational techniques used can significantly impact project team performance. Project managers may exhibit many styles of management throughout the project life cycle.

PMP® & CAPM® Exam

---

5. The X, Y, and Z Factor

Theory X and Theory Y are theories of human motivation created and developed by Douglas McGregor in the 1960's and have been used in human resource management, organizational behavior, and organizational development. They describe two very different attitudes toward workforce motivation. McGregor felt that companies followed either one or the other approach.

Theory Z focused on increasing employee loyalty to the company by providing a job for life with a strong focus on the well-being of the employee, both on and off the job. According to Dr. William Ouchi, its leading proponent, Theory Z management tends to promote stable employment, high productivity, and high employee morale and satisfaction.

Review each situation and determine which style of management is being demonstrated:

The project manager provides opportunity for discussion and participation in decision making	
The project manager spends a great deal of time observing and supervising the project team	
The project team collectively makes decisions	
The project manager directs all project work	
Strong discipline and punishment are used to induce the required work effort	
There is great opportunity for personal improvement and advancement	
An informal administrative control process is utilized	

A. Theory X   B. Theory Y   C. Theory Z

6. Are you motivated? Maslow's Hierarchy of Needs.

Maslow's hierarchy of needs is predetermined in order of importance. It is often depicted as a pyramid consisting of five levels:

Correctly label each level of Maslow's Hierarchy of Needs.

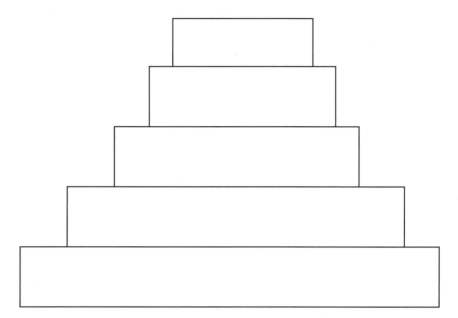

7. More Maslow

Classify each of the following situations by matching the associated level of Maslow's Hierarchy of Needs.

What problems or obstacles may be encountered by the project manager when attempting to staff his or her project team? List the problem or issue and provide a recommendation for managing and resolving the issue.

Situation	Level in Maslow's Hierarchy of Needs
A. A project manager expresses feelings of pride and personal accomplishment	
B. A new project team member requests frequent rest periods, wants the latest labor-saving devices, and expects efficient work methods	
C. A project team member expects to be rewarded publicly for doing his job well	
D. The project team is concerned about the total duration of the project and employment at project completion	
E. The project environment provides an opportunity to work with a team, interact with co-workers, and improve the ability to network with other teams	

8. ✎ **PM Quick Check:** True or false? Abraham Maslow developed the X and Y management theories.

True _____

False _____

Maslow postulated that once a need is satisfied, it loses its motivational importance and is replaced by a different need.

True _____

False _____

Herzberg's Motivator–Hygiene Theory

Frederick Herzberg viewed motivation from a different perspective than Maslow. He believed that certain factors, if present, can result in dissatisfaction among employees and are referred to as *hygiene factors*. The work environment is associated with hygiene factors but the existence of a good work environment does not necessarily result in a motivated work force. In addition to a pleasant work environment, the manager must include *motivating factors* that will encourage employees or project teams to expand their potential and increase their interest in achieving project objectives.

9. Review the list of factors and place them in the appropriate columns.

Hygiene Factor	Motivator

A. Compensation (pay)
B. Sense of responsibility
C. Opportunity for advancement
D. Level of supervision
E. Opportunity for personal growth
F. Peer relationships
G. Company policies and administrative procedures
H. Challenging work
I. Reward and recognition
J. Working conditions
K. Variety in the work
L. A more autonomous environment

PM Knowledge Note

> *The project manager who fails to build and maintain alliances and positive relationships will soon find opposition or indifference to project requirements and objectives.*

☑ **Motivating Techniques:**

- Assigning work that is challenging
- Defining and communicating clear expectations about project work and performance
- Providing appropriate feedback and proper criticism
- Giving honest performance appraisals
- Developing a team attitude toward completion of objectives
- Providing the proper level of direction and support

10. ✎ **PM Quick Check:** True or false?

A. In Maslow's hierarchy of needs, the first level of the hierarchy is associated with hygiene factors such as good working conditions and appropriate levels of supervision.

_____ True

_____ False

PMP® & CAPM® Exam

B. Theory Z managers are authoritarian in style, distrust their employees, and spend great amounts of time observing and directing the work that is being performed.

_____ True

_____ False

11. **Authority**

☑ A project manager's authority is a combination of his or her power and influence. Authority is generally associated with rank or position and granted or delegated by higher level management or one's superiors. Power comes from credibility, expertise, and sound judgment. Failure to establish a project manager's authority may result in:

- Poor communications
- Misleading information
- Antagonism
- Poor working relationships with upper management, peers, and associates
- Surprises from the customer or the project team

PM Knowledge Note

> ***Common sources of power and authority problems:***
> *Poorly documented or no formal authority*
> *Dual accountability of personnel*
> *Two bosses/project managers*
> *Organizational encouragement of individualism*
> *Shifting of personal loyalties from vertical to horizontal lines*
> *Ability to influence or administer rewards and punishment*

Complete the following statement:

The RAM (Responsibility Assignment Matrix) is a tool that:

PMP® & CAPM® Exam

_____

_____

_____

_____

12. Power and Interpersonal Influences

Project managers are frequently required to obtain performance results and achieve project objectives by influencing others to perform activities that they would not normally become involved in; therefore an understanding of interpersonal influences and types of power is critical for a project manager to be successful.

Match each type of interpersonal influence with the correct description.

Type of influence	Description
1. Legitimate power	
2. Reward power	
3. Penalty power	
4. Expert power	
5. Referent power	

A. A person has the ability or is enabled to offer salary increases, promotions, bonuses, and future work assignments.
B. Officially empowered by rank or position in the organization.
C. A person has strong ties to another person of much greater authority and influence. The ability to influence other people due to the challenges and attractiveness of a project.
D. Ability to coerce people through actions that are perceived as resulting in unpleasant or undesirable results. Use of disciplinary actions or the threat of disciplinary action.
E. Ability to influence others through special knowledge, experience, or demonstrated ability.

13. Power-Up

What type of power or interpersonal influence is being used in each of the following situations? A. Formal, B. Reward, C. Penalty, D. Expert, E. Referent.

PMP® & CAPM® Exam

1. As stated by the project sponsor in the project charter, I am the project manager and the team will be expected to take direction from me on any matter that is related to the project.

   _____

2. The project manager is a close friend of the CEO and they live next door to each other. It would be a good idea to do what she asks.

   _____

3. I have two tickets to tomorrow night's playoff game. They are yours if you complete this assignment today.

   _____

4. Either you complete the work assigned to you by the end of the week or the bonus associated with this project will be cancelled.

   _____

5. The speaker is worth listening to. He has a lot of experience and has written several books on the subject.

   _____

14. Team Development

Project managers must develop an environment that encourages teamwork and mutual support. The goal of every team member should be the successful completion of the project deliverables by working together and through effective integration of all project components. Sometimes barriers to team development are experienced. The project manager, as the leader of the team, must work toward the resolution of these barriers to achieve objectives.

For each of the following potential barriers to team development, provide a possible approach or solution that may be taken by a project manager to overcome the barrier:

Barrier	Recommended Solution or Action
Different priorities and interests	
Role conflicts	
Project objectives and expectations are not clear	
Dynamic project environment—Scope changes, new or frequently changes objectives	
Competition over team leadership	
Lack of a defined team structure	
Selection of team personnel	
Credibility of the project manager	
Lack of team member commitment to the project	
Poor project communications	
Lack of sponsor, senior management, or executive-level support	

15. Leadership

☑ Project managers are expected to manage the project by integrating all of the components of the project, communicating information, negotiating for resources and maintaining control of overall project performance. For many project managers this means working with a team of people that may not report directly to the project manager. To accomplish the objectives of the project, the project manager must be perceived by the project team as a strong leader. This includes the development and communication of clear and complete project as well as individual expectations.

PM Knowledge Note	*Project leadership is the ability to get things done well through others. It is an ability to attract followers by providing a clear destination, inspiring high performance, and motivating team members to succeed.*

Roles of the leader and roles of the manager:

Review each of the following roles and align them with either the leader or the manager using the table on the next page. Some roles may be shared by leaders and managers.

A. Plan and budget
B. Motivate and support
C. Organize work groups
D. Inspire team spirit
E. Staff the project
F. Establish a vision
G. Align the team members
H. Control the work flow
I. Short-term view
J. Winning team respect
K. Establishing loyalty
L. Enabling a group to perform as a team
M. Tactical approach
N. Excellent communicator
O. Directs work
P. Sees risks only as threats
Q. Rewarding performance
R. Explaining and selling an idea
S. Setting objectives
T. Focus on the bottom line
U. Encouraging innovation
V. Enabling the team
W. Telling how and when
X. Sees risks as opportunities
Y. Long-range view

Leader	Manager

16. ✑ **PM Quick Check:** Describe the major differences between Maslow's theory and Herzberg's theory about motivation.

Maslow:

Herzberg:

PMP® &
CAPM®
Exam

17. Communications

Project managers must ensure that the right information gets to the right people at the right time and in a cost-effective manner. Effective communication is essential for project success and a process for communicating information should be established in the early stages of the project.

> *Effective communication may include:*
> - *An exchange of information*
> - *An act or instance of transmitting information*
> - *A verbal or written message*
> - *A process by which meanings are exchanged between individuals through a common system of symbols*
> - *A feedback loop to ensure mutual understanding*

Communications Breakdowns

Many project managers experience barriers that impede communications or distort information. These barriers can disrupt project plans, create conflict, and ultimately affect the outcome of the project. It is the responsibility of the project manager to create an environment in which project communications requirements have been aligned properly with the stakeholders involved. Many barriers can be overcome or prevented by clearly stating objectives, establishing roles and responsibilities, and identifying the specific requirements of the stakeholders involved in the project.

A. What are the three basic elements of the communications process?

_____

_____

_____

B. How much time does a project manager typically spend communicating while managing a project?

_____

50%
25%
90%
40%

C. In the space provided create a list of activities a project manager may perform that require some form of communication.

_____

_____

_____

D. Exercise: Label each component of the communications process

Communications process flow:

```
┌────────┐      ┌────────┐      ┌────────┐      ┌────────┐
│        │ ───▶ │        │ ───▶ │        │ ───▶ │        │
│        │      │        │      │        │      │        │
└────────┘      └────────┘      └────────┘      └────────┘
```

> *Typical communications barriers:*
> *Sender credibility*
> *Personality and interests*
> *Attitude and emotion*
> *Position or rank*
> *Existing relationships*
> *Selective listening*
> *Lack of feedback*
> *Evaluative tendency*
> *Assumptions about the sender*

PM Knowledge Note

Communications Channels—A Formula Worth Remembering

PMP® & CAPM® Exam

If you have ever wondered just how many channels of communication exist within a project team, try the formulas $N(N-1)$ divided by two where N equals the number of people on the project team or $N = X(X-1)/2$ where X is the number of people on the team and N equals the number of channels.

18. ✍ **PM Quick Check:** If the number of team members increases from 6 to 12, what is the increase in the number of communications channels?

> Current team member count = 6
> 6 new members are added.
> The number of communications channels has increased by:

PMP® & CAPM® Exam

PM Knowledge Note

> *The communications environment is affected by:*
> *Logistics and geographic issues*
> *Personal contact*
> *Virtual or extended teams*
> *Technology—telephone, electronic data transfer*
> *External factors—politics, economic climate, regulatory agencies*
> *Internal factors—power games, withholding information, indirect communications, blocking, or selective listening*

19. Match the correct description or characteristic with the communication skill or style.

Communication Skill and Personality Type	Matching Letter	Characteristic or Description
1. Authoritarian		**A.** Eager to fight or be disagreeable
2. Promotional		**B.** "Tough guy." Can lower morale through threats
3. Facilitating		**C.** Give expectations and specific guidance
4. Conciliatory		**D.** Not open or outgoing. Does not offer opinions or share information
5. Judicious		**E.** Cultivates team spirit
6. Ethical		**F.** Provides guidance as required, non-interfering
7. Secretive		**G.** Uses experience to arrive at sound decisions
8. Disruptive		**H.** Honest, fair, generally "by the book"
9. Intimidating		**I.** Breaks apart unity of the group, an agitator
10. Combative		**J.** Friendly, agreeable, builds a compatible team

20. Which of the following is a hygienic factor according to Herzberg?

a) Working conditions

b) Reward program

c) Promotion opportunity

d) More challenging assignment

21. Effective Meetings

Project managers can expect to attend many meetings during the project life cycle and may be required to lead many of them. Meetings are actually projects within the project and, depending on the type of meeting and the intended audience, could require a significant amount of planning. Exercise: Create a list of activities that would be required to schedule a meeting and a list of guidelines that should be followed to ensure the meeting is a success.

Meeting Scheduling Activities	Meeting Rules and Guidelines

Kerzner "Quick tips" for the Project Management Institute PMP® and CAPM® EXAM

HUMAN RESOURCES MANAGEMENT AND COMMUNICATIONS MANAGEMENT

Project management is related to many of the functions of general management—planning, organizing, staffing, executing, and controlling.

The communications model involves the region of experience of the sender and the receiver, the encoding and decoding process, personality and perception screen, and a feedback loop.

Feedback is critical for effective communication.

The communications channel formula is $N(N-1)/2$ where N is the number of people on the team.

Communication can be formal, informal, verbal, or written.

Approximately 80% of communications is nonverbal. Maintain an awareness of body language and other nonverbal forms of communication.

Listening is an important part of communication.

Communications barriers will be encountered during the project life cycle. Project managers must maintain an awareness of potential barriers and develop methods to prevent them or resolve them to minimize the potential negative impact on the project.

Every meeting should be considered to be a project. Meetings should be planned appropriately to minimize unproductive time and should always include an agenda . Avoid unnecessary meetings and establish meeting guidelines to ensure greatest productivity and accomplishment of meeting objectives.

Project managers assume both leadership and managerial roles during project planning and implementation.

Managerial styles—Differentiate between Theory X, Theory Y, and Theory Z.

Maslow focused on the Hierarchy of Needs – Physiological, Safety, Social, Esteem, Self Actualization

Herzberg focused on hygiene factors and motivating factors. Hygiene factors include work environment, level of supervision, compensation. Motivating factors include chance for advancement, rewards, challenging work

Personal Influences and Power – Reward power, penalty power, expert power, formal/legitimate power, referent power

RAM – Responsibility Assignment Matrix – aligns task responsibility with specific stakeholders or project team members

Additional tips and practice items for the PMP® exam are included in each chapter and in the section of the workbook entitled **PMP® Exam and PMBOK® Guide Review.**

Answers to Questions and Exercises

1. Planning, organizing, staffing, controlling, directing

2. Measuring—B, Evaluating—C, Correcting—A

3. Project completion and accomplishment of group and organizational objectives.

4. Staffing—D, Training—G, Supervising—A, Delegating—F, Motivating—E, Counseling—C, Coordinating—B

5.

The project manager provides opportunity for discussion and participation in decision making	Y
The project manager spends a great deal of time observing and supervising the project team	X
The project team collectively makes decisions	Z
The project manager directs all project work	X
Strong discipline and punishment are used to induce the required work effort	X
There is great opportunity for personal improvement and advancement	Y
An informal administrative control process is utilized	Z

6. Top—Self-actualization. Next level—Esteem. Next level down—Social. Next Level down—Safety and Security. Bottom or lowest level—Physiological needs/shelter

7. A—Self-actualization, B—Physiological, C—Esteem, D—Safety & Security, E—Social

8. First question –False, Second question - True

9. Hygiene factors—A,D,F,G,J. Motivators—B,C,E,H,I,K,L

10. A—False, B—False

11. RAM—aligns project team members with activities of the WBS

12. 1—B, 2—A, 3—D, 4—E, 5—C

13. 1—A, 2—E, 3—B, 4—C, 5—D

14.

Barrier	Recommended Solution or Action
Different priorities and interests	Establish objectives, goals, and vision. Connect project to personal value
Role conflicts	Use a RAM to establish roles and responsibility
Project objectives and expectations are not clear	Review objectives and revise to meet SMART criteria
Dynamic project environment—Scope changes, new or frequently changed objectives	Establish change control process, clarify requirements, review and revise scopes statement
Competition over team leadership	Obtain sponsor support, establish roles and responsibility early in the project
Lack of a defined team structure	Create an organization chart for the project
Selection of team personnel	Establish staffing requirements, obtain sponsor support
Credibility of the project manager	Demonstrate leadership, display confidence, ask for assistance when it is needed, recognize technical expertise of the team, maintain integrity, gain trust through actions, listen to the team
Lack of team member commitment to the project	Explain importance of the project, connect project to organizational goals, obtain sponsor support
Poor project communications	Identify stakeholder needs, create a communications plan, use a PMIS
Lack of sponsor, senior management, or executive-level support	Establish expectations with management, provide executive-level status reports, involve the sponsor when necessary, connect project to strategic objectives

15. Leader—B, D, F, J, K, L, N, Q, R, S, U, V, X, Y. Manager—A, C, E, G, H, I, J, K, L, M, N, O, P, Q, S, T, W.

16. Maslow—Motivation was a set of clear steps in a hierarchy. As one step is achieved, it is no longer a motivator. Herzberg—Separated hygiene needs from motivation.

17. A—Sender/Encoder, Receiver/Decoder, Feedback
B—90%
C—Developing the scope statement, explaining objectives, preparing status reports, solving conflicts, negotiating for resources, obtaining and reporting status (many other activities may be listed)
D—Sender, Encode, Decode, Receive

18. 6 x (6 − 1) divided by 2 = 15, add 15 (addition of 6 new members). Answer = 30

19. 1—C, 2—E, 3—F, 4—J, 5—G, 6—H, 7—D, 8—I, 9—B, 10—A

20. Which of the following is a hygienic factor according to Herzberg?

a) Working conditions

21.

Meeting Scheduling Activities	Meeting Rules and Guidelines
Set meeting time	Start and end on time
Identify location	Use an agenda
Prepare agenda	Use a parking lot for new issues
Prepare meeting documentation	Do not allow interruptions
Send notice/invitation	Encourage listening

Your Personal Learning Library

Write down your thoughts, ideas, and observations about the material in the chapter that may assist you with your learning experience. Create action items and additional study plans to assist you in enhancing your skills or for preparing to take the PMP® or CAPM® exam.

Insights, key learning points, personal recommendations for additional study, areas for review, application to your work environment, items for further discussion with associates.

_____

_____

_____

_____

_____

_____

_____

_____

_____

Personal Action Items:	
Action Item	**Target Date for Completion**

Chapter Six

MANAGEMENT OF YOUR TIME AND STRESS

Time management includes more than activity duration estimates and developing a project schedule. The demands of the project manager assignment require an organized and disciplined approach to managing personal time. This is due to the often turbulent environment associated with managing a project. There are countless meetings to attend, email to respond to, daily conflicts to resolve, crises to handle, and change requests to analyze. With these factors in mind, effective time management becomes critically important to a project manager's success and well-being.

Time is viewed by many project managers as a constraint and constraints often result in a sense of being rushed or short of time. The perceived improper use of time, the feeling that time is moving too fast, or the failure to manage time effectively can introduce a significant amount of stress into the daily activities of the project manager. It is, therefore, necessary for the project manager to find ways to use time as a resource instead of a constraint and to focus on how to manage and, if possible, reduce the factors that may result in stress.

PM Knowledge Note

Time management requires a well-thought-out and consistent approach to one's daily activities. Here are some recommendations: Prioritize the issues, screen out the issues you should not be involved in, determine your level of involvement in the issues you accept, and do not assume responsibility for a functional manager's issues.

Glossary of terms Key terms and definitions to review and remember

Management by Exception Working within a defined process and identifying and responding to issues that represent a breach of the uniform process. The philosophy of monitoring the operating results of implemented plans and comparing them with expected results.

Milestone A significant event in time. Examples: end of a phase, project completion, completion of a deliverable. Does not have a duration and does not consume resources.

Opportunity Cost   The difference between the yield that funds earn in one use and the yield they could have earned if they were placed in an alternative investment.

Stress   A state of mental or emotional strain or suspense. To test the limits of.

Time Management   The development of processes and tools that increase a business's time efficiency. The ability to manage and control time. Implementing a routine method of scheduling actions that enforce a regimen to fit with a person's flow of work and production activities.

Activities, Questions, and Exercises

Refer to Chapter Six of *Project Management: A Systems Approach to Planning, Scheduling, and Controlling* (10th Edition) for supporting information. Review each of the following questions or exercises and provide the answers in the space provided.

The following questions and exercises are associated with the knowledge areas of the *PMBOK® Guide*: Project Human Resources Management and Project Time Management. Some elements of risk management are also addressed in this chapter.

1. **Time Management Quick Check—A Self-Assessment**
    Review the following questions, provide your response based on your current work environment, and then write a recommendation for managing the issue.

Question	Response based on current environment	Recommendation for managing the issue
Do you have trouble completing work within the allocated time frame or by the established deadline?		
How many interruptions do you experience each day?		
Do you have a procedure for handling interruptions?		
If you need a large block of uninterrupted time, is it available? Is overtime required for this?		
How do you handle drop-in visitors and phone calls?		
How is incoming mail/email managed?		
Do you have established procedures for routine or recurring work?		
Are you accomplishing more or less than you were three months ago?		

Question	Response based on current environment	Recommendation for managing the issue
How difficult is it for you to say no?		
How do you approach detail work?		
Do you perform work that should be handled by your subordinates?		
Do you have sufficient time each day for personal interests?		
Do you still think about your job when away from the office?		
Do you make a list of things to do each day? If yes, is the list prioritized?		
Does your schedule have some degree of flexibility?		
What am I doing that I don't have to do at all?		
What am I doing that can be done better by someone else?		
What am I doing that could be done as well by someone else?		
Am I establishing the right priorities for my activities?		

2. Time Robbers: The Usual Suspects

Consider the following typical time robbers many project managers encounter during the project life cycle. What is your immediate response to each time robber and what can be done to prevent them from occurring?

Time Robber	Immediate Response	Preventive Measure
A job poorly done that must be done over		
Telephone calls, mail, email		
Changes without direct notification/explanation		
Failure to delegate, or unwise delegation		
Too many levels of review		
Too many meetings		

Executive meddling		
Lack of authorization to make decisions		
Company politics		
Poor functional status reporting		
Dealing with unreliable subcontractors		
Desire for perfection		

List other time robbers you have experienced and how you responded to them. What lessons did you learn and how can you manage them more effectively?

Time Robbers—Your personal experience	How you responded	Lessons learned

PM Knowledge Note

Rules and actions for time management:
- *Conduct a time analysis (time log)*
- *Plan solid blocks of time for important things*
- *Classify your activities—important, urgent, not urgent, not important, project related, administrative*
- *Establish priorities*
- *Establish opportunity cost for activities*
- *Train your system (organization)—boss, subordinate, peers*
- *Practice delegation*
- *Practice management by exception*
- *Focus more on opportunities, less on problems*

☑ **Project management Quick Quote: "I have to implement plans I didn't design, but if the project fails, I'm responsible."**

3. Managing Stress

Project managers may encounter several issues that could result in high levels of stress. These issues include but are not limited to: responsibility without authority, a need to be in control, a necessity for perfection, deadlines, role ambiguity, conflicts with or among functional groups, span of control, crossing organizational boundaries, keeping up with fast-paced information. It is important for project managers to maintain awareness about stress and how it can impact job performance and personal health.

Stressed out: Match the following behaviors with the stress symptom.

Stress Symptom	Behavior
A. The hurry pattern	1. Caffeinated drinks, late-night work assignments, making time to get it done
B. Schedule addiction	2. Avoiding big problems, dealing with short-term issues only, waiting until being forced to act
C. Call-waiting syndrome	3. Low tolerance for waiting, complaining when service is too slow, red lights take forever, hang up if put on hold for more than a few seconds
D. No sleep spiral	4. Must stay connected because "they need me"
E. Unfit for the task	5. Often sick, headaches, upset stomach, stiff neck, low endurance
F. Ostrich option	6. Irritable, quick to react harshly to reasonable requests
G. Anger mismanagement	7. Tightly managed time, even for play and with family

Time management issues can be related to risk management. Many of the time robbers listed in this chapter can also be considered potential risks. Effective risk management includes identification of potential risk situations, prioritizations of risks, appropriate response, and monitoring and control of project risks. Remember that each knowledge area is inter-related. Effective planning requires an understanding of how each knowledge area is related to other knowledge areas. Risk management and time management are connected in many ways.

PMP® & CAPM® Exam

PM Knowledge Note

Priorities may change day-to-day. It is a good practice not to postpone what your team can do today.

☑ **Time management techniques:**

- Delegate
- Follow a schedule
- Decide fast whenever possible
- Learn to say no
- Start now
- Do the tough part first
- Travel light
- Work at travel stops and while you are waiting
- Refuse to do unimportant work
- Look ahead
- Ask yourself: Is this trip necessary?
- Ask yourself: Is this meeting necessary?
- Understand and manage your energy cycle
- Rest when you need it

4.   ✎ **PM Quick Check:** True or false?

A. The greater the number of hours worked, the greater the productive output of an individual.

_____ True

_____ False

B. It is always up to the project manager to get it done. If the functional manager can't do it, do it yourself.

_____ True

_____ False

C. You can always find the time later to take care of important matters.

_____ True

_____ False

☑ **More quick quotes:**

For a long time I admired people in a hurry, until I realized that they were merely under stress. What I fear most about stress is not that it kills, but that it prevents one from savoring life.
—Jean-Louis Servan-Schreiber, *The Art of Time*

Rule number 1 is, don't sweat the small stuff. Rule number 2 is, it's all small stuff.
—Dr. Robert S. Elliott, University of Nebraska cardiologist

Lost time is never found again.
—Benjamin Franklin, *Poor Richard's Almanac*

"The bad news is time flies. The good news is you're the pilot."
—Michael Altshuler

"Many people seem to think that success in one area can compensate for failure in other areas. But can it really?...True effectiveness requires balance."
—Stephen Covey

"One worthwhile task carried to a successful conclusion is worth half-a-hundred half-finished tasks."
—Malcolm S. Forbes

PM Knowledge Note

> *Most people spend far too much time reacting to events, and too little time doing things that are really important. Make sure you know the difference.*

☑ **Stress relievers.** There is no simple solution for managing and overcoming stress. Stress reduction depends on the individual and the situation or environment. It is important to identify ways for reducing stress to keep you focused more effectively on the project objectives.

Ideas for stress relief: exercise, satisfaction through accomplishment of tasks, meaningful use of time, taking time to do something enjoyable like working on a hobby, take a walk, or use relaxation techniques several time a day.

5. List some fun or enjoyable activities that you can do to help keep stress under control:

Fun or enjoyable activities that take only a few minutes	Fun activities that take a few hours	Fun activities that take half a day or longer

6. The best way to manage time is to set goals, and to set them using **SMART** criteria. What are SMART objectives?

7. Project Planner – use the following table as a template for creating a time management program to assist you in improving your organizational skills and ability to manage time effectively

Project / Activity	Importance High Medium Low	Urgency High Medium Low	Objectives – What is the desired outcome?	Measures – How will you know you have achieved your objectives?	Target date for completion	Status

Quick tips for managing stress

Perform a personal SWOT analysis.

Identify your personal strengths—Identify what do you do best with respect to people, your job, and other activities you perform. Identify the things you are respected for. Identify your support network including family, co-workers, friends, professional associates.

Identify weak areas—Define your limitations, areas where you feel you are not strong, where you feel people may criticize you. Determine if these weaknesses and criticisms are fair.

Identify opportunities that are available to you—Make a list of opportunities that may be waiting for your action. Determine how your strengths can be used to capitalize on opportunities. What tools can be used to help you with the opportunities?

Determine the consequences of not dealing with your weaknesses. What damage can these threats cause? Identify actions you can take to address and minimize or reduce the threats you have identified.

Learn to see the difference between urgent and important

The important tasks are those that lead you to your goals, and give you most of the long term progress and reward. Those tasks are very often not urgent. Many urgent tasks are not really important. People tend to work on non- urgent, non-important tasks that give the appearance of accomplishing many things but amount to very little. Focus on important items that may not be urgent and those items that are important and urgent.. Examples – an urgent and important task may be a report to an executive that is due in two days. A non-urgent, but important task may be to schedule a training session for a new technology that will be introduced in the next year.

Writing a "to do" list is a simple technique that can increase your productivity by 20 percent or more, It also helps to clear your mind and save you energy and stress.

Kerzner "Quick tips" for the Project Management Institute PMP® and CAPM® EXAM

Time Management

Time management is about your personal use of available time and your own productivity along with the appropriate planning and management of the project schedule. Review the processes of time management in the *PMBOK® Guide*—Activity Definition, Activity Sequencing, Activity Resource Planning, Activity Duration Estimating, Schedule Development, and Schedule Control.

Remember that time management is integrated with other knowledge areas. Effective time management considers risks, contingency planning, team productivity and performance, the scope of the project, the WBS, and many other planning items.

Stress can lead to burnout and burnout leads to lost productivity. The project manager must maintain an awareness of stress-related symptoms that may appear during project implementation. Lowering the stress of the project team as well as a project manager's own stress is part of the project manager's responsibility.

Use a time planner or calendar to manage schedules and avoid conflicts.

Use the WBS to break large assignments or tasks into their component parts.

Take the time upfront during the project start-up to organize and plan the project.

Use the input of the project team to prioritize tasks.

Start with the "big picture" and ensure that vision, goals, and objectives have been clearly defined.

Remain flexible to changes. Change can be expected on just about any project. Instead of resisting change, analyze it and determine if it is beneficial or if it is required.

Additional tips and practice items for the PMP® exam are included in each chapter and in the section of the workbook entitled *PMP® Exam and PMBOK® Guide Review.*

Answers to Questions and Exercises

1. This is an exercise that has many possible answers and is intended to provide you an opportunity to plan possible actions based on your current project environment. Use this exercise as a basis for planning your personal time management processes and to identify priority issues to address.

Example:

Do you have trouble completing work within the allocated time frame or by the established deadline?	Too many projects going on at one time. Dealing with less urgent issues first. Taking on more work than there is time to complete.	Reduce the number of projects, obtain assistance from a less busy project manager, prioritize work items, learn to say no.

2. There are several possible actions that may be taken to address these issues.

Time Robber	Immediate Response	Preventive Measure
A job poorly done that must be done over	Identify areas of poor quality and review with appropriate persons	Set quality expectations, clearly define requirements, determine root causes
Telephone calls, mail, email	Respond to urgent messages only	Schedule time periods for email, phone calls. Reduce volume through better communication and clarity of information
Changes without direct notification/explanation	Respond based on urgency and critical nature of the change. Question the change for validity	Establish a process for communicating change requests
Failure to delegate, or unwise delegation	Determine what can be delegated. Identify appropriate resources to assign work. Prioritize your work and focus on your own responsibilities	Review your work load, define items that can be delegated, develop trust in your employees or project team
Too many levels of review	Escalate, expedite the process	Eliminate unnecessary approval levels
Too many meetings	Attend only the most critical meetings, avoid meetings whenever possible	Make sure the meeting is necessary. Verify that you are needed at the meeting. Improve meeting productivity by establishing controls
Executive meddling	Provide requested information. Inquire about any concerns that may cause executives to meddle in the project work	Establish expectations with executives at project start-up. Define communications requirements

Time Robber	Immediate Response	Preventive Measure
Lack of authorization to make decisions	Obtain authorization through approved channels	Negotiate for increased authority
Company politics	Obtain advice from mentors, follow established practices	Develop strong relationships with key decision makers
Poor functional status reporting	Define immediate status needs	Establish standards for reporting status
Dealing with unreliable subcontractors	Review contractual terms and conditions, escalate as necessary	Establish expectations and performance levels at contract signing. Negotiate penalty clauses
Desire for perfection	Review requirements and emphasize project objectives. Review acceptance criteria and halt any action that has been completed at original criteria	Communicate quality requirements, project budget constraints, expectations about performance and service-level agreements

3. Stressed out A—3, B—7, C—4, D—1, E—5, F—2, G—6

4. A—False, B—False, C—False

5. Use your imagination and list things that you can do to keep stress down. Examples: Take a brief walk, call a friend, go out to lunch, read a book, take a nap, help someone else, exercise.

6. Specific, Measurable, Attainable, Realistic, Time based

7. This table is intended for you to develop plans to improve you use of time. Consider the value of the project or activity as it relates to you available time and the benefits that will be achieved by accomplishing the project and objectives. Consider personal as well as organizational benefits and the risks associated with each project.

Your Personal Learning Library

Write down your thoughts, ideas, and observations about the material in the chapter that may assist you with your learning experience. Create action items and additional study plans to assist you in enhancing your skills or for preparing to take the PMP® or CAPM® exam.

Insights, key learning points, personal recommendations for additional study, areas for review, application to your work environment, items for further discussion with associates.

_____

_____

_____

_____

_____

_____

_____

_____

_____

_____

Personal Action Items:	
Action Item	**Target Date for Completion**

Chapter Seven

CONFLICTS

Conflicts will be encountered in most, if not all, phases of the project life cycle. They may occur at any level, between functional groups and among any of the project stakeholders. Project managers often assume the role of conflict manager and may be required to respond to a wide range of issues. This role is just one of the many roles a project manager may assume during the project.

The ability to handle conflict requires an understanding of why conflicts occur, the sources of the conflict, and the specific needs of the stakeholders involved. In many cases conflict begins with poorly defined objectives and unclear roles and responsibilities.

PM Knowledge Note

Project Objectives: *To avoid conflict, project objectives should be based on the following criteria: specific, not overly complex, measurable, tangible, verifiable, realistic, attainable, established within resource constraints, and consistent with organizational plans, policies, and procedures. These criteria are often summed up using the word SMART—SPECIFIC, MEASURABLE, ATTAINABLE, REALISTIC, and TIME BASED.*

Glossary of terms Key terms and definitions to review and remember

Collaboration   Any cooperative effort between persons or organizations to achieve common goals.

Compromise   Determine a solution where each side leaves with some degree of satisfaction. Generally, a give-and-take process where neither side achieves 100% of its objectives.

Conflict   Conflict is a natural disagreement resulting from individuals or groups that differ in attitudes, beliefs, values or needs. It can also originate from past rivalries and personality differences. Other causes of conflict include trying to negotiate before the timing is right, misunderstanding of information, different priorities, confusion about roles and responsibilities.

Conflict Escalation   The escalation of a conflict to make it more destructive, more confrontational, or otherwise uncomfortable.

Conflict Management   The variety of ways by which people handle grievances, clashes, or right and wrong. It may include avoidance, withdrawal, compromise, forcing, and collaboration.

Conflict Resolution   Conflict analysis, negotiation, facilitation, mediation, arbitration, and judicial settlement. The act of arbitrating differences of belief or opinion about a given state of conditions or circumstances. A process of resolving a dispute or disagreement.

Forcing   Imposing a decision or resolution without opportunity for suggestions or feedback

Needs   Things that are essential to our well-being. Conflicts arise when we ignore others' needs, our own needs or the group's needs. Needs are different than desires or wants.

Smoothing   Removing short-range erratic variations, eliminating jagged edges. In conflict management it is the attempt to give the appearance that progress is being made or that the conflict is not as severe as originally perceived. Smoothing does not generally resolve a conflict.

Values   Beliefs or principles we consider to be very important. Serious conflicts arise when people hold incompatible values or when values are not clear. Conflicts also arise when one party refuses to accept the fact that the other party holds something as a value rather than a preference.

Withdrawing   Avoiding or stepping away from a conflict

Activities, Questions, and Exercises

Refer to Chapter Seven of *Project Management: A Systems Approach to Planning, Scheduling, and Controlling* (10th Edition) for supporting information. Review each of the following questions or exercises and provide the answers in the space provided.

The following questions and exercises are associated with the knowledge areas of the *PMBOK® Guide*: Project Human Resources Management and Project Communications Management. Some elements of Professional and Social Responsibility are also addressed in this chapter.

PM Knowledge Note

Importance of Objectives in Conflict Management

The specific objectives of the project must be communicated to all stakeholders. Failure to provide clearly defined objectives will result in different interpretations at all levels of management and across all organizations involved in the project.

Conflict is not always negative. It can lead to positive outcomes when effectively managed. Healthy conflict can lead to...
- Personal and organizational growth and innovation
- New ways for solving problems and increased creative thinking
- Additional management options

1. ✍ **PM Quick Check:**

A. There are generally three types of objectives associated with a project. List the three primary types of project objectives:

1. _____
2. _____
3. _____

B. Well-written objectives are generally associated with five main criteria. These criteria are often described using a common word as an acrostic (each letter represents the start of another message) . What is the word and what are the criteria? (Example – TEAM = **T**ogether w**E** **A**ccomplish **M**ore)

PM Knowledge Note	*In some cases, objectives may start out very general in nature and are refined or re-established as the project evolves and progresses. This may be associated with progressive elaboration.*

2. Which of the following tools or techniques is specifically used to clearly define the roles and responsibilities of the project team members by aligning responsibility with project activities and tasks and minimizing the potential for conflict between functional groups? Circle the correct answer.

A. WBS
B. RAM
C. Activity List

3. Exercise: Reasons for Conflict. List at least 10 common sources of conflict. For additional learning value, provide possible solutions for or preventive measures that can be taken to avoid or minimize the conflict.

Sources of Conflict	Bonus Section—Solutions

4. ✍ PM Quick Check:

Conflict is a risk factor that always produces a negative result.

True_____   False_____

Conflict always results in an unpleasant work environment and leads to more serious issues that cannot be resolved.

True _____   False

The most effective method for resolving conflicts is to compromise.

True _____   False

5. Exercise: What would you do?

Conflicts can occur with anyone over anything. How would you manage the following conflicts?

A. Two functional managers on your team appear to be having personality clashes and almost always assume opposite points of view during decision-making discussions. They are both from the same functional organization.

_____

_____

_____

B. The manufacturing department reports that it cannot produce the end product according to engineering specifications.

_____

_____

_____

6. Conflict is not always associated with negative results and damage. Explain why conflict may be beneficial to a project team.

_____

_____

PM Knowledge Note

Many executives feel that the best way of resolving conflicts is by establishing priorities.

7. Fill in the blank:

The connection between the organizational goals and project priorities is established by the _____ and the project manager.

A. Project Team
B. Customer
C. Sponsor
D. Stockholders

PMP® & CAPM® Exam

8. Project Conflict Intensity

During the project life cycle, a project team will go through several stages of team development. As the project life cycle progresses, the level of conflict generally decreases in its intensity. Name the four stages of team development and explain how conflict intensity changes in each phase.

Team Development Phase	Conflict Intensity

9. ✎ **PM Quick Check:**

A. The less specific the objectives of a project, the more likely that conflict will develop.

True _____ False _____

PMP® & CAPM® Exam

B. When conflict occurs during project implementation the project managers should:

a) Study the problem and collect all available information
b) Develop a situational approach or methodology
c) Set the appropriate atmosphere or climate
d) All of the above

C. List the 5 major conflict resolution modes. Which of the modes will actually resolve the conflict?

Conflict-Handling Mode	Solves Conflict? Yes or No

10. Power, Influence, and Conflict

How can each of the following be used to prevent, reduce, or resolve conflict within the project environment?

A. Authority or position of power	
B. Penalty power	
C. Work challenges	
D. Promotion/advancement opportunity	
E. Theory X style management	
F. Theory Y style management	
G. Communications plan	
H. RAM	

11. The Conflict Meeting

During the project life cycle, it may become necessary for the project manager to schedule a conflict or confrontation meeting between conflicting parties to find a solution. Review the steps in the process and match them with the correct description or explanation.

Steps or events in the conflict process	Matching Number	Description of the step or event
A. Setting the climate		1. Obtaining feedback on the implementation
B. Analyzing the images		2. Obtaining cross-organizational involvement, obtaining commitment
C. Collecting the information		3. Taking action on the plan
D. Defining the problem		4. Forming cross-functional problem-solving groups
E. Sharing the information		5. Obtaining input to be used to resolve the problem
F. Setting the appropriate priorities		6. Establishing a willingness to participate
G. Organizing the group		7. How do you see others? How do they see you?
H. Problem solving		8. Identifying and clarifying all positions
I. Developing the action plan		9. Making the information available to all participants
J. Implementing the work		10. Getting the feelings and concerns out in the open
K. Following up		11. Developing work sessions and time tables

12. Match the Mode Description

Review each of the following situations and select the correct corresponding conflict-handling mode in use. Place a check mark in the appropriate boxes.

PMP® & CAPM® Exam

Situation	Withdrawal	Smoothing	Forcing	Compromise	Collaborate/ Confront
A. The requirements are my decision and we are doing it my way					
B. I've thought about it and you are right; we'll do it your way					

C. Let's discuss the alternatives. Perhaps there are alternatives					
D. Let me again explain why we need the new requirements					
E. See my section supervisors; they are handling it now					
F. I've looked over the problems and I might be able to ease up on some of the requirements					

☑ **Conflict Management in Practice**

Conflict Analysis Exercise: Use this approach to identify, understand, and resolve conflicts.

Answer these questions to help develop an approach to resolving a conflict.

- Who are the groups involved?
- Why are they involved?
- How are they organized?
- Are the groups capable of working together?
- What is the history between the groups?
- How did the conflict develop?
- What are the main issues? What are the secondary issues?
- How can negative issues be reframed into more positive statements?
- How negotiable are the positions that have been taken?
- What are the common issues?
- What values and interests have been challenged?
- What external constraints or other influences must be considered?
- What past positive experiences can be used to influence cooperation?
- What is the timeline or urgency of the conflict and the need for a solution?
- How would an outside negotiator be received?
- Is an external negotiator needed?

Conflict Analysis Table – Use this table to assist in assessing conflicts that may be experienced during the project life cycle. Consider how the conflict originated, who is involved and their specific needs as well as their motivators.

People or Groups Involved Consider: Who do they represent? Who are the leaders? Cultural issues. Sensitivities	Conflict Description	Reason for the conflict – background information, history, incidents leading to the conflict Primary and secondary issues. What values and or interests are affected? What is negotiable?	Strategy for Resolution

Kerzner "Quick tips" for the Project Management Institute PMP® and CAPM® EXAM

Human Resources Management and Communications Management

Remember the five types of conflict-handling modes — Withdrawing, Smoothing, Compromise, Forcing, and Collaboration.

There are many sources or causes of conflict. The most frequent causes are: schedules, project priorities (especially in a matrix environment), resources, technical opinions, administrative procedures, cost and cost estimates, personalities.

Remember the four phases of team development — Forming, Storming, Norming, and Performing.

The five types of power and influence are associated with conflict management: expert, reward, penalty, legitimate or formal, and referent power.

Managerial styles: Theory X, Theory Y, and Theory Z may impact the level of intensity of conflicts and the frequency of occurrence.

Negotiation is a key factor in conflict management.

A project communication plan and a project management information system may reduce the probability of conflict.

Clearly defined and communicated objectives are a key element in reducing and preventing conflict.

Additional tips and practice items for the PMP® exam are included in each chapter and in the section of the workbook entitled *PMP® Exam and PMBOK Guide® Review.*

Answers to Questions and Exercises

1. A. Ultimate project objective, phase or deliverable objectives, task and activity objectives
   B. SMART — Specific, Measurable, Attainable, Realistic, Time-based

2. RAM

3. Resources (people), equipment, facilities, capital expenditures, cost estimates, technical opinions, trade-off decisions, priorities, administrative procedures, scheduling, assignment of responsibilities, personality clashes

4. False   False   False

5. A. There are several possible answers to this question. Set up a conflict resolution meeting, identify the specific conflicts, identify the needs of each side, search for common ground.
   B. Review the requirements. Identify where the problems exist. Arrange for a meeting between engineering and production. Identify alternatives. Decide on the appropriate course of action. Reach agreement.

6. Conflict surfaces problems as well as opportunities. Brainstorming solutions can be a team-building exercise and can generate new ideas and opportunities.

7. Sponsor

8.

Team Development Phase	Conflict Intensity
Form	Low but will increase during this phase
Storm	Highest — Team members deal with uncertainty, roles, responsibilities
Norm	Decreasing as work is performed
Perform	Lowest — Conflict is not eliminated but in this phase the team is working well together and can resolve most conflicts without project manager intervention

9.
   A. True
   B. D

C.

Conflict-Handling Mode	Solves Conflict? Yes or No
Withdrawal	No
Smoothing	No
Compromise	Yes — This is a win-lose/lose-win approach but it does solve the conflict to some extent
Forcing	Yes — This approach solves the immediate problem but causes other conflicts
Collaboration	Yes — Best approach to a win-win solution

10.

A. Authority or position of power	Eliminates debate
B. Penalty power	Forces resolution through fear of punishment
C. Work challenges	Creates environment that encourages teamwork
D. Promotion/advancement opportunity	Provides incentive to work as a team
E. Theory X style management	Related to forcing. No participative decision making
F. Theory Y style management	Participative style. Encourages collaboration
G. Communications plan	Effective distribution of information can reduce the potential for conflict
H. RAM	Clear understanding of responsibilities often reduces conflict

11.

A — 6
B — 7
C — 10
D — 8
E — 9
F — 11
G — 4
H — 2
I — 5
J — 3
K — 1

12.

Situation	Withdrawal	Smoothing	Forcing	Compromise	Collaborate/Confront
A. The requirements are my decision and we are doing it my way			X		
B. I've thought about it and you are right; we'll do it your way	X				
C. Let's discuss the alternatives. Perhaps there are alternatives					X
D. Let me again explain why we need the new requirements		X			
E. See my section supervisors, they are handling it now	X				
F. I've looked over the problems and I might be able to ease up on some of the requirements				X	

Your Personal Learning Library

Write down your thoughts, ideas, and observations about the material in the chapter that may assist you with your learning experience. Create action items and additional study plans to assist you in enhancing your skills or for preparing to take the PMP® or CAPM® exam.

Insights, key learning points, personal recommendations for additional study, areas for review, application to your work environment, items for further discussion with associates.

_____

_____

_____

_____

_____

_____

_____

_____

_____

_____

Personal Action Items:	
Action Item	**Target Date for Completion**

Chapter Eight

SPECIAL TOPICS

There are several situations and topics that require a project manager's attention. The project manager's main focus of attention is the successful completion of the project but the type of project, the diversity of the project team, the types of personal interfaces, and the overall project environment may require the project manager to address and manage many if not all of the following items:

- *Performance measurement*
- *Compensation and rewards*
- *Managing small projects*
- *Managing mega projects*
- *Morality, ethics, and the corporate culture*
- *Internal partnerships*
- *External partnerships*
- *Training and education*
- *Integrated project teams*
- *Virtual project teams*

Each of these item or situations may be included in the project manager's job description or assigned as a responsibility during project planning and implementation.

PM Knowledge Note

Benefits of performance measurement and appraisal systems:
- Ensures that the right people are in the right positions.
- Identifies areas issues that require attention before they become unmanageable
- Appraisals provide an objective approach to measure levels of individual and team accomplishment and to determine appropriate compensation.
- Identifies team and individual employee training requirements.
-Provides project managers with the data needed to determine if employees have potential for advancement.

Glossary of terms   Key terms and definitions to review and remember

Competency   The ability to transfer and apply skills and knowledge to situations and environments. The knowledge and skill and the application of that knowledge and skill to the standard of performance required in employment.

Conflict of Interest   A situation where an individual is placed in a compromising position where the individual can gain something of personal value or enrichment based on the decisions made.

Expectation   A result that is considered most likely to happen. A belief that is centered on the future,

Mega Project   A large undertaking with multiple components and technologies that requires a significant amount of resources over a long period of time and at substantial cost and risk.

Performance Appraisal   A method by which the ability of an employee to complete job assignments (generally in terms of quality, quantity, cost and time) and other assigned responsibilities is evaluated and reported to the employee through feedback . Performance appraisal is a part of career development.

Performance Measurement   The process of developing measurable indicators that can be systematically tracked to assess progress made in achieving predetermined goals. Determining performance gaps between what customers and stakeholders expect and what each process produces in terms of quality, time, and cost.

Performance Standard   An objective performance level that must be met. A statement of expectations or requirements established by a supervisor or manager for a performance element at a particular rating level.

Professional Responsibility   A mastery of a special body of advanced knowledge that bears directly on the well-being of others. The special moral responsibilities and requirements that constrain professionals to apply their knowledge in ways that benefit society.

Activities, Questions, and Exercises

Refer to Chapter Eight of *Project Management: A Systems Approach to Planning, Scheduling, and Controlling* (10th Edition) for supporting information. Review each of the following questions or exercises and provide the answers in the space provided.

The following questions and exercises are associated with the Knowledge Areas of the PMBOK® Guide: Project Human Resources Management and Project Communications Management. Some elements of Professional and Social Responsibility are also addressed in this chapter.

PM Knowledge Note	*The employee performance appraisal process allows the project manager to identify, evaluate, and prepare developmental actions to improve poor performance and to motivate an employee to continue to reach for higher levels of performance.*

1. **Measuring Project Team Performance.**

   During project execution what areas of team performance should be evaluated by the project manager?

   _____

   _____

2. ✍ **PM Quick Check:** In which structure will the project manager have the greatest influence in the preparation of an employee's performance appraisal?

   A. Functional
   B. Matrix
   C. Projectized

PM Knowledge Note	*360-degree feedback*—A process in which a manager obtains feedback, generally anonymously, from peers, subordinates, supervisor, and customer. This process assists in defining strengths and areas for improvement in interpersonal and managerial skills.

3. During which stage of the project life cycle should the project manager establish performance appraisal criteria?

   _____

4. What actions can a project manager take to increase functional employee loyalty and commitment to the project?

   _____

   _____

   _____

5. Besides monetary compensation, what can a project manager do to motivate his/her project team?

   _____

   _____

   _____

PM Knowledge Note	*The foundations of compensation practices are based on four systems:* *Job classification* *Base pay* *Performance appraisals* *Merit increases*

6. ✎ **PM Quick Check:**

   A. Defining job titles and corresponding responsibilities are critical to project success and should be acted upon in the early stages of project planning.

   True _____   False _____

   PMP® & CAPM® Exam

   B. According to Herzberg, compensation is a motivating factor.

   True _____   False _____

7. Why are proper financial compensation and rewards important to the morale and motivation of the project team?

   _____

   _____

8. What is the purpose of the employee performance appraisal?

   _____

   _____

9. Generally, organizations measure the performance of their project managers in what two major areas?

   _____

   _____

10. Project managers are measured on their performance in specific areas. These areas are viewed in terms of primary and secondary importance. Review each measure and identify if it is a primary or secondary measure of performance.

Measure	Primary	Secondary
Target costs		
Overhead reduction		
Low team conflict		
Key milestones		
Quality		
Technical accomplishments		
Project staffing		
Budget Development & Control		
Performance measures and controls		
Reports and reviews		
Inter-functional communication		
Responsiveness to changes		

11. The project team, functional managers, and project resources are usually assessed on their abilities associated with:

   A. Technical implementation
   B. Team performance
   C. A & B

PM Knowledge Note

If done well, the appraisal should provide particular measures of job performance that assess the level and magnitude at which the individual has contributed to the success of the project including managerial performance and team performance components.

☑ **Study notes:** Common challenges or issues that may be encountered by a project manager in a small company or when working on small projects:

   • The project manager may have to assume several responsibilities and act as a functional manager and a project manager
   • The project manager may have to manage multiple projects
   • There may be limited resources, especially in a matrix environment
   • There is a significant need for strong interpersonal skills
   • There are generally shorter lines of communication
   • In small companies there is usually no PMO
   • In small companies there is a greater risk to the total company if the project fails
   • There may be greater financial constraints and tight control of funding in small organizations
   • Upper management may be more likely to interfere with or become directly involved in day-to-day project activities
   • Estimating activities are usually required to be more precise in smaller companies

12. Whether managing a small project or a mega project the project manager may encounter similar issues. Review the issues listed in the table. Explain strategies for dealing with these issues.

Situation	Strategy
Multiple projects	
Limited resources	
Lack of trained or competent resources	
Internal competition for materials	

13. During project planning and implementation a project manager may be required to develop strategies to meet project objectives. Review each strategy and explain the advantages and disadvantages of each strategy.

Strategy	Advantage	Disadvantage
Assign the best employees to the most visible or highest priority projects		
Use of overtime to resolve schedule slippages		
Adding additional staff if the project slips		
Hiring subcontractors when internal resources are not available		
Assigning a technical expert as the project manager		
Establishing a projectized organizational structure		
Utilizing a balanced matrix structure		

☑ **Study notes: Professional and Social Responsibility, Morality, Ethics, and Corporate Culture**

Professional responsibility is generally the mastery of a special body of advanced knowledge that bears directly on the well being of others. The special moral responsibilities and requirements constrain professionals to apply their knowledge in ways that benefit society. Project managers must maintain sensitivity to cultural, ethical, political, and geographical factors when planning a project and managing a team. There are factors both internally and externally that may create significant challenges for the project manager and impact how decisions are made.

Internally driven adversity—being asked by your management to take action that is in the interest of your company but violates your own moral and ethical beliefs:

- You are asked to lie to the customer in a proposal in order to win a contract
- You are asked to withhold bad news from your management
- You are asked to withhold bad news from the customer
- You are ordered by management to violate ethical accounting practices to make your numbers look good to senior management
- You are instructed to ship a potentially defective unit to a customer to meet production quotas
- You are asked to cover up acts of embezzlement or charge the wrong account codes
- You are asked to violate the confidentiality of a private personal decision by a team member

Externally driven adversity—you are asked by your customer to take action that may be in the customer's best interest but violates your own moral and ethical beliefs:

- You are asked to destroy information that could be damaging to the customer during legal action taken against the customer
- You are asked to lie to consumers to help maintain the customer's public image
- You are asked to release unreliable information that could be potentially damaging to a customer's competitor

Tasks associated with Project Management Professional and Social Responsibility

- Ensure individual integrity and professionalism
- Contribute to the project management knowledge base
- Enhance individual competence
- Balance stakeholder interests

14. What would you do? Review each situation and select a response that most closely represents your viewpoint.

   A. An assistant project manager had the opportunity to be promoted and manage a new large project that was about to begin. She needed the project manager's permission to accept the new assignment, but if she left the project manager would have to perform her work in addition to his own for about three months.

   a) Deny the promotion due to the work load
   b) Delay the promotion until a suitable replacement can be found and trained
   c) Approve the promotion and develop a transition process for the assistant project manager
   d) Convince the assistant project manager that the promotion is not really a good opportunity

   B. In the first month of a twelve-month project the project manager discovers that the end date was too optimistic and cannot be met.

   a) Hold the information and wait for a miracle to happen
   b) Change the scope statement to reduce the work required
   c) Advise the customer and the sponsor about the issue and provide alternatives
   d) Continue with the project as planned and arrange for a new project manager to take over as soon as possible

15. ✍ **PM Quick Check: Review each statement and select the best word or phrase in parentheses that completes the statement most appropriately.**

   1. A good project manager will explain the individual and team performance appraisal process (immediately/after the first major milestone).
   2. Proper (financial compensation/rewards to improve morale and stimulate) are the greatest motivators for project teams.
   3. The first step in job classification and creating job descriptions is to (define job titles/establish base pay).
   4. Traditionally, the purpose of the performance appraisal is to (provide justification for salary treatment/prepare for future goal setting).
   5. The performance of the project manager is measured through (technical implementation and task performance/business results and managerial ability).
   6. Companies that promote morality and ethics usually create (internal adversity/a supportive working environment).

16. **Internal and External Partnerships**

   Project managers depend on functional groups to provide resources that are competent and to provide support to accomplish project objectives. Establishing partnerships with internal and external organizations and developing good working relationships will assist the project manager in obtaining the best available resources and creating a willingness to cooperate. Describe actions a project manager may take to establish partnerships with internal functional groups or external organizations such as subcontractors and customers.

Potential Partnership	Possible Action or Actions
Internal functional groups	
Subcontractors	
Customer	
Sponsor	

17. Training and Education

Project managers must continually assess the needs of their project teams as well as their own personal needs for continued self-development. There are three general opportunities for training to occur: on-the-job training, formal education, and knowledge transfer.

To ensure that project objectives are achieved, what activities should a project manager perform to assess the capabilities of each team member and to determine the training needs of the project team?

_____

_____

18. From the project manager perspective, describe an *Integrated project team:*

_____

_____

19. Virtual teams

In today's project environment the project manager can expect to manage project teams that are dispersed geographically due to the nature of the project. These dispersed teams, also known as virtual teams present unique challenges to the project manager. Consider each of the following virtual team situations and develop a potential solution. There may be several potential solutions to each situation.

Situation	Solution
Time zone differences make it difficult to organize team conference calls	
Team members have different perceptions of the project scope	

Cultural differences are creating conflicts among some team members	
Technology differences make it difficult to keep team members updated and informed	

Kerzner "Quick tips" for the Project Management Institute PMP® and CAPM® EXAM

Maintain awareness that professional responsibility is embedded in each knowledge area of the *PMBOK® Guide*.

Review the PMI® Code of Ethics and Professional Conduct for information about ethics, guidelines, and requirements that professional project managers are expected to honor. The purpose of the Code is to instill confidence in the project management profession and to help an individual become a better practitioner.

The PMI® Code of Ethics and professional Conduct focuses on: Responsibility, Respect, Fairness, and Honesty

Your personal integrity is the key factor in professional responsibility.

Performance appraisals, just like project reviews, should be conducted at specific intervals to measure progress and identify areas for improvement.

Become familiar with your company's policies regarding diversity, culture, and ethics.

Review performance measurement criteria and expectations with your project team at the start of the project. Communicate objectives early and provide feedback on a regular basis.

Review project constraints and assumptions during project status meetings and scheduled performance reviews.

Creating internal partnerships with functional managers is an effective way of ensuring that project objectives will be accomplished. Maintain an awareness of functional manager priorities and constraints and establish positive working relationships.

Additional tips and practice items for the PMP® exam are included in each chapter and in the section of the workbook entitled *PMP® Exam and PMBOK® Guide Review.*

Answers to Questions and Exercises

1. Technical judgment, work planning, ability to communicate, attitude toward the team, cooperation, work habits, timeliness, quality of work, on-time completion, adherence to standards.

2. C

3. Earliest stage or planning stage (at start-up or kickoff).

4. Connect the project to organizational goals, create a sense of urgency and esprit de corps, obtain and ensure an understanding of their needs, listen and act on concerns.

5. Recognize accomplishments, offer thanks for work well done, prepare formal recognition and communicate to management, respect the team, trust the team.

6. A. True, B. False

7. Proper compensation should be established to provide a foundation for motivation. Reward and recognition of work improves morale and willingness to continue to perform at higher levels.

8. Assess the employee's work performance against pre-established objectives. Provide justification for salary actions, establish new goals and objectives for the next review period, identify and manage work-related problems, serve as a basis for career discussions.

9. Business results [ROI, profit, on-time delivery, within budget, according to requirements, and managerial ability (effectiveness, leadership, direction, team performance)].

10.

Measure	Primary	Secondary
Target costs	X	
Overhead reduction		X
Low team conflict		X
Key milestones	X	
Quality	X	
Technical accomplishments	X	
Project staffing		X
Budget	X	
Performance measures and controls	X	
Reports and reviews		X
Interfunctional communication		X
Responsiveness to changes		X

11. 3—C

12.

Situation	Strategy
Multiple projects	Use of portfolio management. Prioritize all projects. Focus on most critical projects first. Obtain additional support from the sponsor(s). Transfer some projects to other project managers
Limited resources	Identify critical tasks, schedule resources to meet critical path dates, obtain additional resources externally, escalate to sponsor
Lack of trained or competent resources	Outsource/subcontract, schedule training, establish specific staffing criteria, negotiate with functional managers for more qualified resources
Internal competition for materials	Portfolio management, prioritization, escalation, identify alternate sources

13.

Strategy	Advantage	Disadvantage
Assign the best employees to the most visible or highest priority projects	Project should meet objectives, high quality, less supervision, low maintenance of the team	Other projects do not receive the benefit of the expertise
Use of overtime to resolve schedule slippages	Short-term solution, may return project to acceptable status	Increased cost, potential staff burn-out
Adding additional staff if the project slips	May improve project performance, additional resources for use on the project	Increased cost, learning curve of the new resources, risk of not resolving the problem
Hiring subcontractors when internal resources are not available	Useful contingency, trained resources, possibly short ramp-up or learning curve	Increased cost, potential for lower quality, less/no direct control, loss of expertise at project completion
Assigning a technical expert as the project manager	Understands the technical issues, respected by the team for expert knowledge Less risk of missing details	Too much focus on technical issues, too little focus on integration activities
Establishing a projectized organizational structure	Project manager has high level of authority, controls all resources, streamlined decision making	Must obtain resources from functional groups, may not obtain critical resources when needed, resources are concerned about next assignment

Strategy	Advantage	Disadvantage
Utilizing a balanced matrix structure	Project manager and functional manager have equal authority Potentially better coordination across functional lines	Potential conflict about authority between functional manager and project manager. Resources generally report to the functional manager, functional manager has greater influence on resources

14. A – C, B – C

15. 1—immediately, 2—rewards to improve morale and stimulate, 3—define job titles, 4—provide justification for salary treatment, 5—business results and managerial ability, 6—a supportive working environment

16.

Potential Partnership	Possible Action or Actions
Internal functional groups	Obtain information about their needs first, avoid "telling" them what they have to do, empathize
Subcontractors	Establish clear expectations, tie performance to progress payments, discuss opportunities for future business
Customer	Establish expectations, define customer responsibilities, obtain customer input and approval of plans, involve the customer in planning exercises
Sponsor	Establish expectations, define communications needs, agree on level of support, involve the sponsor as needed

17. Obtain information about the competency level of the team members, review past performance, obtain appraisal information from functional managers, interview team members to determine abilities, ask for input about training needs, assess work output early in the project.

18. The Integrated project team (IPT) includes the supplier's project team, the clients project team, and the required consultants and industry specialists required to achieve project objectives. The IPT brings together the designers, construction resources, maintenance resources, and other entities that will exchange information or provide deliverables in support of the project's product.

19.

Situation	Solution
Time zone differences make it difficult to organize team conference calls	Alternate meetings time to balance the needs of the team members
Team members have different perceptions of the project scope	Schedule a project kick off meeting (in person if possible) to review project objectives and project scope of work.
Cultural differences are creating conflicts among some team members	Review the various cultures associated with the team, investigate areas of potential conflict. Set expectations based on the needs of the project and the client. Seek additional support and advice from resources who are trained in this area. Research the conflicts to gain an understanding before taking action.
Technology differences make it difficult to keep team members updated and informed	Establish technology standards at the start of the project. Look for common applications. Research technology that is available with the organization. Create standards for communicating information

Your Personal Learning Library

Write down your thoughts, ideas, and observations about the material in the chapter that may assist you with your learning experience. Create action items and additional study plans to assist you in enhancing your skills or for preparing to take the PMP® or CAPM® exam.

Insights, key learning points, personal recommendations for additional study, areas for review, application to your work environment, items for further discussion with associates.

Personal Action Items:	
Action Item	**Target Date for Completion**

Chapter Nine

THE VARIABLES FOR SUCCESS

Project success for many managers and project teams refers to the elements of the Triple Constraint—Time (Schedule), Cost (Budget), and Scope (Quality and Performance Specifications). Success is actually more than completing the project within the triple constraint. Success depends on a systems approach to the project where all project elements, planning components, and deliverables of the project are inter-related and managed with an understanding that failure in one area will have an effect on many other areas of the project.

Establishing and communicating success criteria are two of the key responsibilities of the project manager. During the planning process, the project manager provides the team with the information needed to prepare the project plan and establish the controls to keep the project on track to achieve client satisfaction and project success. The variables for project success include:

- *Setting expectations*
- *Planning—including the relationships with the triple constraint*
- *Predicting project success—through analysis and performance measurement*
- *Project management effectiveness*

Project success in many ways depends on the actions of three key stakeholder groups: the project manager and team, the project sponsor or parent organization, and the customer or client organization—the group that will ultimately receive the final project deliverable. Assuring project success requires the following actions:

- *Selecting the "right" people for the project team*
- *Developing commitment and a sense of "urgency" to complete the project successfully*
- *Obtaining the appropriate level of project manager authority and management support*
- *Establishing a good relationship with the client, sponsor, and team members (including functional managers)*
- *Creating a positive public image about the project*
- *Establishing a participative relationship with team members for decision making and problem solving*
- *Developing realistic project objectives for cost, schedule, and performance*
- *Anticipating problems and establishing risk management plans and back-up strategies*
- *Creating an appropriate team structure that is flexible, supportive, and has minimal managerial levels to enhance communication*
- *Maintaining an awareness of project team needs and establishing a strategy for motivation*
- *Maintaining focus on the project objectives and the final end product*
- *Establishing a change control process and communicating the importance of managing change*
- *Developing a strategy for rewarding and recognizing the project team*

**PM
Knowledge
Note**

A project cannot be successful unless it is recognized as a project by upper level management. Without upper management support the project manager will not be able to obtain the resources required to achieve project objectives. Project success is also associated with the value that is achieved at project completion. In some cases, true project value may not be realized until many years after the project is completed.

Glossary of terms   Key project management terms and definitions to review and remember

Best Practice   Something that works well on a repetitive basis. Something that leads to a competitive advantage. Something that provides value to both the buyer and seller. A process, technique, or action that has been proven to produce superior results consistently and can be applied enterprisewide.

Deliverable   A tangible, verifiable work output. (Note: Not all deliverables on projects go to the customer; some deliverables are just for the PM.)

Expectation   Something that is considered reasonable, due, necessary, or obligated.

Functional Manager   Generally a subject matter expert or technical expert from a specific functional group such as engineering, design, or other technological areas who may control or manage resources required to meet project objectives. A person with management authority over a specific organizational unit that performs a specialized function or produces a product.

Lessons Learned   Information obtained through experience that may be shared to improve performance, increase opportunity, or prevent occurrence of negative events, results, or activities.

Project Sponsor   The person or organization that has authorized the project and provides the funding to support the project. (Note: Some organizations feel that it is best for the ultimate sponsor to come from an area that does not have a vested interest in the project such that impartial decisions may be made, such as termination for failing to provide a valid business case.)

Stakeholder   Any person or organization directly involved in or impacted either positively or negatively as a result of the project.

Triple Constraint   The triple constraint consists of three major project elements, Schedule, Cost, and Performance Specifications (scope and quality), usually depicted as a triangle and indicating the competing demands associated with a project. The triple constraint emphasizes the relationship between each of the elements and the impact of a change to any side of the triangle. Most project trade-off decisions are based on the relationships with the triple constraint:

> **Cost/Budget**   The estimated cost to deliver the product or service

> **Schedule**   The estimated time line for the project

> **Scope/performance Specifications**   Also refers to quality

Value   An amount, as of goods, services, or money, considered to be a fair and suitable equivalent for something else; a fair price or return. Monetary or material worth. Worth in usefulness or importance to the possessor; utility or merit. In the project management environment the perception of value is associated with the views of the project stakeholders and their specific definitions of value. Value may be seen in terms of achieving the objectives associated with the triple constraints or it may also be defined in terms of return on investment, level of customer satisfaction, and benefits that may be realized in the future well after the project has been completed.

Activities, Questions, and Exercises

Refer to Chapter Nine of *Project Management: A Systems Approach to Planning, Scheduling, and Controlling* (10th Edition) for supporting information. Review each of the following questions or exercises and provide the answers in the space provided.

The following questions and exercises are associated with the knowledge areas of the *PMBOK® Guide*: Integration Management, Scope Management, Human Resources Management.

1. List five principal factors that are commonly used to measure and determine if a project is successful.

a) _____

b) _____

c) _____

d) _____

e) _____

2. Explain how each of the following organizational variables can have an impact on project success:

Organizational Variable	Effect on the project
a) Strategic planning	
b) Enthusiastic management support	
c) Prompt and accurate communication from upper-level management	
d) Organizational structure flexibility	
e) Emphasis on past experience (lessons learned)	

3. Achieving Project Success

There are many methods available and actions that may be taken to achieve project success. Review the following statements and complete each statement by selecting the appropriate phrase from the list provided:

Projects are more likely to be successful if you:

1. Encourage openness and honesty from _____

2. Create an atmosphere that _____

3. Plan for _____

4. Develop short and informal _____

5. Avoid excessive _____

6. Allow adequate time to _____

7. Ensure that project work packages are _____

8. Match the right people with_____

9. Develop effective working relationships with _____

10. Establish and use effective_____

11. Connect project assignments and task responsibility with _____

12. Plan for project completion and transfer of final deliverables _____

    a. Adequate funding to complete the entire project

    b. The start of the project

    c. Encourages teamwork and healthy competition

    d. Lines of communication

    e. Paper work and administrative workload

    f. Establish the project groundwork or foundation and define the project work

    g. The proper size, are manageable, and have been assigned organizational responsibility

    h. Performance appraisal and rewards

    i. The right jobs and with the appropriate level of training and competency

    j. The project team, functional managers, and project sponsor

    k. During project execution and before actual project completion

    l. Planning and control procedures

PM Knowledge Note

After completion and acceptance of deliverables, many projects require a period of time (one to two months or more) after the work is completed for administrative reporting, final cost summary, and hand-off to the receiving organization. An acceptance period may be established within the contractual terms and conditions of the project.

4. Major Causes of Project Failure

Project managers concentrate on the successful completion of a project but should maintain an awareness of why projects may fail. There are several reasons for project failure but the most common are:

- Selection of a concept that is not applicable or does not have a sound basis
- Forcing a change at an inappropriate time
- Selecting the wrong person as project manager—assigning a person who possesses significant technical skills but does not have sufficient managerial and interpersonal skills
- Executive or upper management is not supportive of the project—lack of support from management will, in most cases, create significant roadblocks and eventually result in project cancellation
- Inadequately defined tasks—a system for planning that includes task definition, development of a schedule, estimated costs, and control processes is required or the project team will experience considerable amounts of rework, delays, and additional costs
- Management techniques—a tendency to do more than what is required, failure to build teamwork, lack of motivation, oversupervision, and poor communications skills will contribute to project failure
- Unplanned project termination—This may be caused by a change in strategic direction, loss of funding, uncontrolled costs, poor quality, lack of team commitment, or contract default

Exercise: For each of the potential reasons for project failure provide a possible solution or recommended preventive action.

Reason for Failure	Preventive action
1. Concept is not applicable or is unsound	
2. Forcing a change at the wrong time	
3. Selecting the wrong person as project manager	
4. Management is not supportive of the project	
5. Inadequately defined tasks	
6. Ineffective management techniques	
7. Unplanned project termination: a) change in strategic direction b) loss of funding c) uncontrolled costs and other variances d) poor communication	

5. Project Management Effectiveness

Project success and a successful career path are dependent upon the project manager's ability to establish and maintain good working relationships with management as well as the project team. Clear expectations between the project manager and all key project stakeholders must be established early in the project planning process. There are four key variables in measuring the effectiveness of the project manager in dealing with stakeholders: Credibility, Priority, Accessibility, and Visibility.

Exercise: Review the list of characteristics and match them with the key variable of management effectiveness.

Project Management Effectiveness—Key Variables	Characteristic of effectiveness
Credibility	
Prioritization	
Accessibility	
Visibility	

a) sound decision maker
b) "selling" the importance of the project and connecting it with organizational goals
c) experience from a variety of assignments
d) making a positive impression when presenting the project to upper management
e) emphasizing facts over opinions
f) ensuring that others receive appropriate credit for work performed
g) obtain testimonial support for the project from functional groups, customers, and other managers
h) ability to communicate as needed with stakeholders
i) become personally known by managers in several departments including upper management
j) create an environment where managers and customers desire your involvement
k) conduct timely informational meetings
l) use available publicity media appropriately
m) create a common understanding about the project through presentations and emphasize project successes

> *Generally, a project manager is expected to:*
> *Assume total accountability for project success or failure, provide timely and accurate project reports and information, minimize disruption of organizational operations during project execution, manage interpersonal issues among project team members, maintain a self-starting attitude and approach, and grow in experience with each assignment.*

PMP® & CAPM® Exam

6. Managing Expectations

Project success is dependent not only on the abilities of the project manager but also on a set of clearly defined expectations that have been established between the project manager and the project sponsor or executive, between the project manager and project team members, and between the project manager and the customer.

The following items are typical expectations that are established during the beginning stages of the project life cycle. Review each item and determine if it is an expectation set between the project manager and project sponsor/upper management or between the project manager and project team.

Expectation	Project team expectation of the project manager	Sponsor/executive expectation of the project manager	Project manager expectation of the project team
Take action on requests			
Provide clearly defined decision channels			
Assist in problem solving			
Provide direction and leadership			
Assist in conflict resolution			
Provide feedback			
Facilitate interaction between functional groups			
Provide strategic direction			
Define expectations clearly			
Demonstrate innovation and creativity			
Stimulate team building and group process			

Expectation	Project team expectation of the project manager	Sponsor/executive expectation of the project manager	Project manager expectation of the project team
Facilitate the addition of new team members			
Protect the team from external pressures and politics			
Remain results oriented			
Communicate effectively and timely			
Fully commit to the project objectives			
Reward and recognition for the work accomplished			
Respect			
Motivate and maintain high morale			

7. The Importance of Lessons Learned

Lessons can be learned during any phase of a project. To make them useful they should be documented and shared with other project managers and project teams within your organization. Sharing lessons learned is associated with the Professional and Social Responsibility domain of Project Management. Transfer of knowledge, increasing the project management historical data base of your organization, and enhancing the processes you work with are all associated with task 2 of the Project Management Professional Role Delineation Study published by the Project Management Institute.

PMP® & CAPM® Exam

✍ **PM Quick Check:** True or false?

1. Lessons learned should only be reviewed and documented after the project has been completed.

  _____ True

  _____ False

2. Project success will be assured if the project team maintains control of the scope, schedule, and cost of the project.

  _____ True

  _____ False

8. **Best Practices**

Most organizations, especially those which are considered "project based," have experienced the advantages of documenting and sharing lessons learned. Examples of lessons learned include:

- Establishing project life cycle phases
- Standard methodology for planning
- Use of templates for planning, scheduling, control, and risk management
- Control of customer and contractor-generated changes
- Use of earned value measurement

It is important to note that what is considered a best practice in one organization may not work effectively in another organization. The project manager and project team may be required to adapt industry best practices to meet their organization's specific needs.

Describe at least one best practice for each of the following items that would be useful to share with project team members and other project managers:

Project Scope	Define the scope as completely as possible uisn previous projects, experience, templates and the inout of all key stakeholders. Explain the reason for the project and who will be affected by the project.
Stakeholders	Identify all project stakeholders. Ensure that all team members understand the definition of a stakeholder. Establish communications plans to meet the needs of the stakeholders. Become aware of the specific bias's of each stakeholder
Standards	Communicate standards to the project team. provide appropriate documentation about applicable standards. Establish control procedures to ensure that the project conforms to standards. Identify exceptions to organizational standards.

Risk management	Establish a risk management plan. Conduct a SWOT analysis. Provide risk management training to the project team. Schedule periodic risk reviews
Estimating time and cost	Use reliable sources for all estimates. Avoid padding. Consider risk when estimating.

9. Explain what is meant by "Critical Success Factors and provide examples.

PM Knowledge Note	*Best practices are those actions or activities undertaken by the company or individuals that lead to a sustained competitive advantage in project management.*

Kerzner "Quick tips" for the Project Management Institute PMP® and CAPM® EXAM

The topic "Variables for Success" is most closely related to the following knowledge areas in the *PMBOK® Guide*: Understanding the Project Environment, Interpersonal Skills, Project Life Cycle and Organization, Project Management Processes, Integration Management

The most common factors used to determine project success are the elements of the Triple Constraint—Time, Cost, and Scope (Scope is also associated with quality and performance specification).

Project success is not based solely on managing the triple constraint. Customer satisfaction, minimizing substantial changes, team satisfaction, and minimizing the impact of the project on business operations are also key factors.

There are five major process groups that are integrated in the planning of a project: Initiating, Planning, Executing, Monitoring and Controlling, and Closing.

The project manager works in collaboration with the project team to determine what processes are appropriate and at what level of effort to achieve project success.

Utilize lessons learned and project failures to identify problem areas and develop preventive measures.

Preventive measures are generally less costly than repair, rework, and failure.

A project review at the end of a phase or at the completion of a milestone will assist in identifying areas for improvement and increasing the probability for overall project success.

Review stakeholder needs to ensure that project performance is on track with expectations.

Review project planning assumptions and constraints on a regular basis. Validate assumptions and maintain an awareness of how constraints may impact project performance.

Additional tips and practice items for the PMP® exam are included in each chapter and in the section of the workbook entitled *PMP® Exam and PMBOK Guide® Review.*

Answers to Questions and Exercises

1.

    a. on time
    b. within budget
    c. according to scope/specifications
    d. quality at the required level
    e. customer satisfaction

2. Effect on the project:

    a. Strategic planning drives the selection of projects
    b. The project will be supported by the entire organization and result in greater buy-in of the project team
    c. Regular feedback and communication from management indicates involvement and continued interest in the project
    d. Enables resources to be moved as necessary to support the project
    e. More efficient planning including better risk management, avoidance of repeated errors, and improved probability of success

3.

    1. b
    2. c
    3. a
    4. d
    5. e
    6. f
    7. g
    8. i
    9. j
    10. l
    11. h
    12. k

4.

1. Perform proof of concept or feasibility study. Obtain additional information. Review requirements and specifications.
2. Utilize a change control process and analyze the impact of the change before deciding to implement.
3. Review the complexity of the project, the visibility of the project, and establish performance expectations for project results and managing people. Establish specific qualifications for the project manager.
4. Review the project charter, verify project objectives, determine if the project is related to organizational objectives, review project risks. Obtain feedback from management.
5. Ensure the WBS has been properly defined to the appropriate level of detail. Verify the scope statement. Conduct a project kickoff and planning meeting.
6. Establish performance criteria for the project manager, obtain a coach or mentor, escalate to the next level of management, provide appropriate training.
7. a—A change in strategic direction may result in project termination. The project manager may not be able to prevent this from occurring. Effective and regular communication with decision makers is recommended.   b—Similar to the situation described in (a). Loss of funding may be an organizational issue. Ensuring that the project objectives remain connected with organizational objectives, ensuring the project is viewed as a priority project, and controlling project results to show positive project progress may prevent termination.   c—Establish project monitoring and control procedures and initiate corrective action before variances approach established thresholds.   d—Establish a communications plan at project start-up. Emphasize the importance of teamwork and integration of project components. Identify stakeholder needs and act on those needs.

5.

Project Management Effectiveness—Key Variables	Characteristic of effectiveness
Credibility	A,C,E,F,J,M
Prioritization	B,G
Accessibility	H,I,J
Visibility	D,I,J,K,L,M

6.

Expectation	Project team expectation of the project manager	Sponsor/executive expectation of the project manager	Project manager expectation of the project team
Take action on requests	X	X	X
Provide clearly defined decision channels	X	X	
Assist in problem solving	X	X	X
Provide direction and leadership	X	X	
Assist in conflict resolution	X	X	
Provide feedback	X	X	X
Facilitate interaction between functional groups	X	X	
Provide strategic direction		X	
Define expectations clearly	X	X	X
Demonstrate innovation and creativity	X	X	X
Stimulate team building and group process	X	X	
Facilitate the addition of new team members	X	X	X

Expectation	Project team expectation of the project manager	Sponsor/executive expectation of the project manager	Project manager expectation of the project team
Protect the team from external pressures and politics	X		
Remain results oriented		X	
Communicate effectively and timely	X	X	X
Fully commit to the project objectives		X	X
Reward and recognition for the work accomplished	X	X	
Respect	X	X	X
Motivate and maintain high morale	X	X	

7. ✐ PM Quick Check

   1. False
   2. False

8.

Project Scope	Define the scope as completely as possible uisn previous projects, experience, templates and the inout of all key stakeholders. Explain the reason for the project and who will be affected by the project.
Stakeholders	Identify all project stakeholders. Ensure that all team members understand the definition of a stakeholder. Establish communications plans to meet the needs of the stakeholders. Become aware of the specific bias's of each stakeholder

Standards	Communicate standards to the project team. provide appropriate documentation about applicable standards. Establish control procedures to ensure that the project conforms to standards. Identify exceptions to organizational standards.
Risk management	Establish a risk management plan. Conduct a SWOT analysis. Provide risk management training to the project team. Schedule periodic risk reviews
Estimating time and cost	Use reliable sources for all estimates. Avoid padding. Consider risk when estimating.

9. Critical Success Factor - any of the aspects of a business that are identified as vital for desired targets to be reached and maintained. Critical success factors are normally identified in such areas as production processes, employee and organization skills, functions, techniques, and technologies. The identification and strengthening of such factors may be similar.

Examples of Critical Success factors

1. Training and education
2. Quality performance data and reporting
3. Management commitment
4. Customer satisfaction
5. organizational structure
6. Quality assurance process
7. Effective Communication
8. Processes for continuous improvement

Your Personal Learning Library

Write down your thoughts, ideas, and observations about the material in the chapter that may assist you with your learning experience. Create action items and additional study plans to assist you in enhancing your skills or for preparing to take the PMP® or CAPM® exam.

Insights, key learning points, personal recommendations for additional study, areas for review, application to your work environment, items for further discussion with associates.

_____

_____

_____

_____

_____

_____

_____

_____

_____

_____

Personal Action Items:	
Action Item	**Target Date for Completion**

Chapter Ten

WORKING WITH EXECUTIVES

Project managers are tasked with many responsibilities during the life cycle of a project. Establishing a positive relationship with executives and continually interfacing with them as the project is planned and implemented are major factors in achieving project success and securing additional project assignments. The role of the executive in the process of managing projects varies significantly from one organization to another but generally the project executive is referred to as the project sponsor. The project sponsor is involved in the project selection process (determining which projects are approved for implementation) and provides or authorizes the necessary funding and resources to achieve project objectives. The project sponsor's role includes providing guidance about: project objectives, setting priorities, creating or approving the project organizational structure, initial upfront planning and conflict resolution. In the early stages of the project the project sponsor assists the project manager in several activities:

- *Establishing the appropriate project objectives*
- *Aligning the project with strategic organizational goals*
- *Providing information about organizational environmental factors and political issues that may impact the project*
- *Justification and prioritization of the project within the performing organization*
- *Providing guidance regarding organizational policies and procedures*
- *Establishing the appropriate level of executive management—client contact and support*

PM Knowledge Note

When establishing a relationship with the project sponsor, the project manager should become familiar with what are described as organizational process assets (policies, procedures, and standards) and enterprise environmental factors (organizational culture) in The PMBOK® Guide— Fourth edition. Knowledge of these items may assist in defining expectations and developing strategies for managing project issues that may require sponsor involvement.

During the initiation or kickoff phase of the project the project sponsor assumes an active role in setting objectives and providing guidance to the project manager. At this time the project sponsor establishes project priorities in both business and technical terms. During the execution phases of the project the sponsor provides support on an as-needed basis and takes a more passive role. Day-to-day activities are managed by the project manager and the sponsor becomes involved when necessary to assist in resolving issues that are beyond the authority level of the project manager. To avoid the possibility of micromanagement (also known as executive meddling) the project manager must establish clear expectations with the sponsor at the start of the project and communicate effectively throughout the project life cycle by providing project progress, status, and forecast reports regularly.

Glossary of terms   Key terms and definitions to review and remember

Enterprise Environmental Factors   Factors that have been established over time or influence an organization's general operating practices. These factors include the organization's recognized culture, government or industry standards, existing human resources, administrative procedures for hiring or performance reviews, organization risk tolerances, and commercial databases used for estimating costs.

Exit Champion/Sponsor   This is an individual or committee at the senior-most levels of management that periodically performs a health check on projects to make sure that the project should continue. Project sponsors do everything possible to make sure that their project continues. The champion determines whether or not the resources can better serve the company elsewhere, whether the enterprise environmental factors have changed, whether the assumptions are still valid and whether the perceived value will still be there at project completion. Champions validate the continuation or termination of the project and generally can override the project sponsor and cancel a project.

Objective   A planned or intended outcome

Organizational Process Assets   Formal and informal processes established and in use within an organization. These may include standard policies for safety, product quality, communication technology, financial controls, change control, and risk management.

Project Sponsor   The person or organization that has authorized the project and provides the funding to support the project. However, some organizations feel that it is best for the ultimate sponsor to come from an area that does not have a vested interest in the project so that impartial decisions may be made, such as termination for failing to provide a valid business case.

Scope Creep   Unauthorized or uncontrolled changes to the project scope. Scope creep sometimes occurs when a project team or project manager attempts to exceed customer expectations through additional work that is not included in the approved scope of work.

SWOT Analysis   A strategic planning method used to evaluate the Strengths, Weaknesses, Opportunities, and Threats involved in a project or in a business venture. It involves specifying the objective of the business venture or project and identifying the internal and external factors that are favorable and unfavorable to achieving that objective

Activities, Questions, and Exercises

Refer to Chapter Ten of *Project Management: A Systems Approach to Planning, Scheduling, and Controlling* (10th Edition) for supporting information and assistance in completing each exercise.

The following questions and exercises are associated with the knowledge areas of the PMBOK® Guide: The Project Management Framework, Project Life Cycle and Organization,

Projects and Strategic Planning, Understanding the Project Environment, General Management Knowledge and Skills, Project Stakeholders, Organizational Influences, Planning Process Group, Project Human Resource Management, Project Communications Management.

Review each of the following questions or exercises and provide the answers in the space provided.

1. **Executive Knowledge: True or False**

   a) Executives should be involved in all phases of project planning and implementation.

      True _____   False _____

   b) The project sponsor provides assistance "as needed" during the execution of the project.

      True _____   False _____

   c) Expectations should be mutually established between the project sponsor and project manager at the start of the project.

      True _____   False _____

   d) The project executive or executive steering committee is responsible for establishing project priority within the organization.

      True _____   False _____

   e) Every project requires a formally identified sponsor.

      True _____   False _____

   f) The project sponsor should be visible to the entire team as well as the customer and constantly informed about project status.

      True _____   False _____

   g) The goals and objectives of the sponsor should be aligned with the strategic goals and objectives of the organization supporting the project and should clearly define the definition of project success.

      True _____   False _____

2. **Role Identification.** Review each role and identify if it is associated with the project manager or project sponsor. Place a checkmark in the appropriate columns.

Role	Project Manager	Project Sponsor
1. Major decision making role and participation in the sales effort and contract negotiation		
2. Prepares detailed project reports		

Role	Project Manager	Project Sponsor
3. Provides assistance in getting the project underway by communicating processes and procedures and securing appropriate staffing		
4. Liaison to executive committees and steering committees		
5. Prepares the detailed project budget estimate		
6. Prepares performance appraisals for project team members		
7. Interprets company policies and communicates information about organizational process assets		
8. Becomes involved in solving major project related problems		

3. Managing the Micromanager

The most effective method for managing a project sponsor or executive who closely manages, observes, and influences day-to-day project manager activities is:

a) Provide the sponsor with large amounts of administrative work to prevent the sponsor from becoming involved in the project activities.

b) Ask for role clarification and establish expectations for providing project information and managing project activities.

c) Invite the project sponsor to every project status meeting and have all team members provide status reports directly to the project sponsor.

PM Knowledge Note

The project sponsor works with the project manager to establish objectives, set priorities, create the project organization, communicate organizational policies, establish client-executive contact, obtain key staffing, establish monitoring and control processes for use during execution, and develop processes for escalation and conflict resolution.

4. Red, Green, and Yellow

Project-based organizations often refer to project status using what is known as the "traffic light" reporting process or red, yellow, and green status condition.

Green light or green status	Work is progressing as planned. No major issues. Sponsor involvement is not required.
Yellow light or yellow status	A potential problem exists. The sponsor is informed but sponsor action is not required at this time.
Red light or red status	A problem exists that may affect the project scope, schedule, or budget. Sponsor involvement and action is required.

Categorize the status of each of the following situations as green, yellow, or red light.

Situation	Status Type (Green, Yellow, Red)
1. A task that is not on the critical path is delayed. There is sufficient slack to manage the delay.	
2. A critical resource may not be available at the scheduled time. Negotiations are underway to resolve the issues.	
3. A report shows project performance regarding scope and schedule, and budget is within acceptable variance thresholds.	
4. The customer insists on a major requirements change well after acceptance had been obtained and production is in advanced stages.	
5. The project manager discovers that a major component of the scope of work is missing in the project plan due to an oversight by the customer.	
6. The client has issued a change request that could delay the completion of a contractual milestone.	
7. A functional manager advises the project manager that overtime may be needed to complete a task on time. A contingency had been included in the cost estimate for the task.	

Making the grade! A PMP exam study tip.

> ***The Status Reporting Process.***
> *Generally there are three types of status reporting—progress reports (what has been accomplished), status reports (the current state of the project), and forecasts (what is expected to occur). Project managers must identify specifically what information the project sponsor expects to receive and when.*

PMP® & CAPM® Exam tip

5. Managing Scope Creep

To avoid scope creep the project manager should:

PMP® & CAPM® Exam

a) Issue an authorization to change the scope each time the customer submits a requirements change request.

b) Ignore all change requests and review them at the completion of the project to determine if a new project should be initiated.

c) Conduct periodic project reviews with the project sponsor to compare project baselines with actual work.

d) Prepare, sign, and deliver to the sponsor a statement that specifically commits to the sponsor that there will be no changes to the project plan once it has been approved.

6. Why is a SWOT analysis important to the strategic planning process?

7. What conditions would cause a project executive or sponsor to become directly involved in the day to day activities of a project?

PM Knowledge Note

> *Occasionally a project manager may disagree with a project sponsor regarding decisions about the project. Disagreements may occur due to the project sponsor's lack of understanding about technical issues, the project sponsor's workload or span of control which may limit available time to support the project, or the communications gaps that may be formed in the organization's hierarchal structure. To avoid disagreements the project sponsor and project manager must establish clear expectations about roles and responsibilities and define how the project is related to organizational strategic goals and objectives.*

Kerzner "Quick tips" for the Project Management Institute PMP® and CAPM® EXAM

The topic "Working With Executives" is most closely related to the following areas in the *PMBOK® Guide*: Human Resources Management, Communications Management, Project Scope Management, and Integration Management.

The project sponsor is the person, group, or organization that provides the financial resources for the project.

Project sponsors may be executives within an organization or they may be steering committees or groups of decision makers. Project managers are usually selected by the project sponsor.

Project managers should set expectations with the project sponsor at the start of the project and revisit those expectations frequently during the project life cycle.

Executives and project sponsors generally require higher level summary information about project status and progress. Milestone charts and high-level summary Gantt charts are usually appropriate methods of communicating status to executives and sponsors.

Project management is related to many of the functions of general management—planning, organizing, staffing, executing, and controlling.

Before escalating a problem to a project sponsor, the project manager should verify that resolution actually requires sponsor intervention.

When escalating an issue to a project sponsor it is important for the project manager to provide possible solutions or alternatives for the sponsor to review.

The project sponsor is considered to be a *key stakeholder*. The project managers must be familiar with the process of *managing stakeholders*, which requires an understanding of communications requirements and the ability to resolve issues that may cause stakeholder dissatisfaction or conflict. Balancing stakeholder needs is also associated with professional and social responsibility.

Additional tips and practice items for the PMP® exam are included in each chapter and in the section of the workbook entitled *PMP® Exam and PMBOK® Guide Review.*

Answers to Questions and Exercises

1.

        a—False
        b—True
        c—True
        d—True
        e—False
        f—True
        g—True

2.

Role #1	Answer
1	Sponsor
2	Project manager
3	Sponsor and project manager
4	Sponsor
5	Project manager
6	Project manager
7	Sponsor
8	Sponsor

3. b

4.

Situation	Condition
1	Green
2	Yellow
3	Green
4	Yellow
5	Red
6	Red
7	Green

5. c

6. The SWOT analysis (Strengths, Weaknesses, Opportunities, Threats) provide the sponsor and the organization with useful information that will assist in the decision to move forward with the project and also assess the probability of project success. The SWOT analysis may also generate alternatives to the original project solution.

7. Failure to report status in a timely manner. Inaccurate or late project status information. Lack of useful information. Failure to maintain adequate levels of communication with the project executive or sponsor. Failure to meet agreed upon expectations.

Your Personal Learning Library

Write down your thoughts, ideas, and observations about the material in the chapter that may assist you with your learning experience. Create action items and additional study plans to assist you in enhancing your skills or for preparing to take the PMP® or CAPM® exam.

Insights, key learning points, personal recommendations for additional study, areas for review, application to your work environment, items for further discussion with associates.

_____

_____

_____

_____

_____

_____

_____

_____

_____

_____

Personal Action Items:

Action Item	Target Date for Completion

Chapter Eleven

PLANNING

Planning can best be described as the function of selecting the enterprise objectives and establishing the policies, procedures, and programs necessary for achieving them. Planning in the project environment may be described as establishing a predetermined course of action within a forecasted environment. The project manager is the key to successful project planning. Ideally the project manager is assigned to the project at the conceptual stage and remains committed until project completion. Planning requires a systematic, flexible approach that will address unique activities, involve the entire project team, and emphasize the need for integration of all plan components. Project planning involves multifunctional input and discipline through controls and reviews. It is an iterative process and continues throughout the life of the project.

A major objective of project planning is to define completely (or as completely as possible) all work required to produce the desired result.

PM Knowledge Note

In the project environment:
If a task is well understood prior to being performed, much of the work can be preplanned.
If a task is not understood, then during the actual task execution more knowledge is gained that in turn may lead to changes in resource allocation, schedules, and priorities.
The more uncertain the task, the greater the amount of information that must be processed in order to assure effective performance of the project team members assigned to the task.

Project Management "Quotable quotes"

"Failure to plan means planning to fail."

"The primary benefit of not planning is that failure will then come as a complete surprise rather than being preceded by periods of worry and depression."

"In preparing for battle I have always found that plans are useless, but planning is indispensable."—Dwight D. Eisenhower

Planning—The Basic Six

There are six basic reasons for project planning:

- To eliminate or reduce uncertainty
- To improve efficiency of the operation
- To obtain a better understanding of objectives
- To provide a basis for monitoring and controlling work
- To make sure that the project's objectives and deliverables are aligned with corporate or strategic objectives
- To provide some degree of assuredness that the perceived value can be obtained at project completion

Planning is a continuous process and is intended to assist the project managers and project team, including the project sponsor, in making decisions with a focus on the desired outcome of the project. Effective planning will improve the efficiency of the team and increase the probability of meeting objectives. It is important to note that a project plan created by the project team based on experience, lessons learned, expert judgment, and an effective process can be expected to change during execution but the plan provides the initial guidance required to organize resources and establishes direction and purpose for the project team.

Planning is also defined in terms of strategic, tactical, and operational perspectives. Strategic planning generally refers to a longer term view of three to five years, tactical planning may include a one- to two-year view, and operational planning is a shorter term view from the current point (now) to six months to a year. When planning a project the project manager and team should consider the relationship of the project objectives to the higher level strategic objectives of the organization. A connection between project objectives and strategic objectives will, in most cases, generate sustained upper management support.

General planning includes:

- Setting objectives
- Relationship to a higher level program or business operation
- A schedule
- A budget
- Forecasting—a projection and estimate of the desired results and other factors that may impact objectives
- An organization—the duties and responsibilities associated with achieving objectives
- Policy—guidelines for decision making
- Procedures—the method for carrying out policies and plans
- Standards—guidelines for acceptable performance

Planning is actually a system that includes setting objectives, developing work descriptions, developing a network diagram to identify dependencies, scheduling tasks, defining activities at the appropriate levels of detail, budgeting based on task identification and resource requirements, tracking performance and reporting progress, providing feedback, managing change, and eventually closing out the project.

Glossary of terms   Key terms and definitions to review and remember

Assumptions   For planning purposes, items that are believed to be true, real, or certain.

Configuration Management   A process to manage changes to products through surveillance. Configuration management ensures that changes to physical and functional characteristics of the products or deliverables of the project are reviewed and approved. Configuration management also includes the establishment of an archived audit trail of all changes, approved and not approved, as well as updates and a history of all baseline changes.

Constraints   Boundaries or limitations that may affect planning such as contractual dates, project funding, resource availability.

Objective   An aim or end of an action. Something toward which effort is applied. A purpose to be attained. Generally, objectives are described in specific terms, can be measured, and are attainable and action oriented, realistic, and bound by time.

Plan   A formal and approved document that defines how the project will be implemented. It guides project execution, documents planning assumptions, facilitates communication among stakeholders, and provides a baseline for measuring project performance. The project plan is expected to change through the project life cycle. The amount of change depends upon the complexity of the project, the level of risk and uncertainty, the economic environment, and the completeness of the planning process.

Portfolio   A collection of projects or programs and other work that are grouped together to facilitate effective management of that work to meet strategic business objectives.

Procedure   A series of steps followed in a definite order to accomplish a desired objective or result.

Process   A set of interrelated actions and activities designed to bring about a specific result.

Program   A group of projects managed in a coordinated way to obtain benefits and control not available from managing them individually.

Progressive Elaboration   Continually moving forward in increments, adding greater levels of detail.

Project   A temporary endeavor (has a start and finish date) undertaken to create a unique product or service, and most likely involving multifunctional disciplines.

Project Charter   The document that authorizes the existence of the project and the use of organizational resources. Usually issued by the project sponsor or project initiator.

Project Life Cycle   Generally, a collection of sequential project phases that defines the work and duration of the project. (However, based upon the risk accepted, project life cycle phases can overlap.)

Project Phase   A discrete component or element of a project that includes specific work activities to be completed in a scheduled time frame. Project phases are bound by specific start and end dates. The project life cycle is composed of a series of project phases.

Project Scope   The work that must be performed to deliver the product, service, or desired result of the project.

Rolling Wave Planning   Rolling wave planning is the process of planning for a project in a progressive manner where work to be accomplished in the near term is planned in detail and work that is planned for the future is planned at a higher level and elaborated later as the project continues through the life cycle.

Scope   The sum of the products and services to be provided as a project.

Scope Statement   Description of the major deliverables, project objectives, and project assumptions. The scope statement provides a basis for making future decisions about the project.

Standard   A document established by consensus that provides for common and repeated use. Generally considered as rules or guidelines to be followed in an organization.

Statement of Work   A narrative description of work to be done under contract.

Subsidiary Plan   A subset of the project management plan. Subsidiary plans are documents that provide detailed guidance and support for the project management plan and are developed based on the needs of the project. Examples: Scope Management Plan, Cost Management Plan, Risk Management Plan, Communications Plan.

Work Breakdown Structure   A deliverable-oriented hierarchal decomposition of the work to be executed by the project team. The WBS organizes and defines the total scope of the project.

Work Breakdown Structure Dictionary   A document that describes the components of the work breakdown structure in detail

Activities, Questions, and Exercises

Refer to Chapter Eleven of *Project Management: A Systems Approach to Planning, Scheduling, and Controlling* (Tenth Edition) for supporting information and assistance in completing each exercise.

The following questions and exercises are most closely associated with the knowledge areas of the *PMBOK® Guide*: The Project Management Framework, Project Life Cycle and Organization, Projects and Strategic Planning, Understanding the Project Environment, General Management Knowledge and Skills, Project Integration Management, and Project Scope Management.

Review each of the following questions or exercises and provide the answers in the space provided. Some questions and situations will have more than one appropriate answer or response. Project managers are aware that many issues they face on a daily basis may have several possible solutions. These exercises are designed to encourage you to think about alternatives before taking action.

1. **The Logic of Planning**

    To be truly effective, planning must include:

    *   Agreement about the purpose and objectives of the project
        *   Agreement about the stakeholders associated with the project
        *   Input from the key stakeholders
        *   Management support
    *   Assignment and acceptance of individual responsibilities within the project team and among key stakeholders
    *   Coordination of work activities
    *   Commitment of the team to the group/project goals and objectives
    *   Lateral or cross-organizational communication

Key questions to consider during the planning process: Match the question to the planning process.

Planning process step	Questions
1. Prepare environmental analysis	
2. Set objectives	
3. List alternative strategies	
4. List threats and opportunities	
5. Prepare forecasts	
6. Select strategy portfolio	
7. Prepare action programs	
8. Monitor and control	

a) Where are we now? How did we get here?
b) Where are we capable of going? What do we need to take us where we want to go?
c) Are we on course? If not, why not? What do we need to do to be on course? Can we do it?
d) What do we need to do? When do we need to do it? How will we do it? Who will do it?
e) Is this where we want to be? Where would we like to be? In one year? In five years?
f) What might prevent us from getting there? What might help us get there?
g) Where will we go if we continue as before? Is that where we want to go? How could we get to where we want to go?
h) What is the best course for us to take? What are the potential benefits? What are the risks?

2. ✑ **PM Quick Check:** True or false?

PMP® & CAPM® Exam

1. Assumptions should be validated regularly during the project life cycle.

_____ True

_____ False

2. The project manager should develop all of the detailed plans for the functional managers to prevent omissions and inconsistency.

_____ True

_____ False

3. Project managers should welcome top management participation, especially at the start of the project.

_____ True

_____ False

3. Life Cycle Phases

An effective approach to planning is to divide the project into specific phases and creating a project life cycle. Applying a project life cycle approach provides consistency within an organization by defining key phase deliverables and also establishes a greater ability to plan and control the activities in each phase. Analyzing the results or accomplishments of each phase assists the project manager and sponsor in determining if the project should continue on to the succeeding phase. These "end of phase" reviews are commonly referred to as phase exits, stage gates, or go and no-go decision points. End of phase reviews assist in determining the "health" of a project and can prevent an organization from continuing with a project that will not meet financial goals, achieve the desired value, or use resources that could be deployed more effectively and beneficially on other projects.

Some key points about project life cycles and project phases

PMP® & CAPM® Exam

- All projects go through a life cycle that has specific and describable phases
- Each phase is initiated, planned, executed, monitored, controlled, and brought to closure
- Phases generally have specific products or intended accomplishments that, upon completion, provide the starting point for the next phase
- The costs at the start of the project life cycle are low and increase throughout the project
- Transitions from one phase to another provide an ideal time to conduct reviews to compare performance with baselines
- Phases can overlap to compress the total project schedule if the risks are considered acceptable. This technique is referred to as "fast tracking"

Commonly used life cycle phases:

- Conceptual
- Feasibility
- Preliminary planning
- Detailed planning
- Execution
- Testing and commissioning

-or-

Initiate

Plan

Execute

Evaluate

Complete the following statement

The benefits of developing and using a standard project life cycle are:

_____

✎ **PM Quick Check:**

The purpose of a feasibility study is to:

a) Determine the total cost of the project and establish a baseline budget.
b) Provide technical data and alternatives to the concept, and form a basis on which to decide whether to undertake the project.
c) Determine the return on investment before initiating the project.

Projects are divided into phases for which of the following reasons?

a)   To improve the ability to control project activities and resources and manage deliverables more effectively

b)   To establish gates for project reviews

c)   To provide consistency for managing projects within an enterprise

PM Knowledge Note	*Providing interim deliverables on projects that may have long life cycles will give the customer and sponsor a sense that work is being accomplished. Interim deliverables may indicate a series of successes that can assist in maintaining team focus and commitment toward the project.*

4. Project Manager and Functional Manager Roles

The responsibilities of the project manager and the functional managers assigned to the project are specific and unique. Match each of the following roles with either the project manager or the functional manager.

Responsibility	Project Manager	Functional Manager
Set project goals and objectives		
Develop detailed task descriptions		
Communicate/Establish major milestones		
Set high-level requirements		
Provide detailed schedules		
Identify areas of risk, uncertainty, and conflict		

Responsibility	Project Manager	Functional Manager
Establish ground rules and provide assumptions		
Define project constraints		
Provide summary status reports for the sponsor		

5. Project Planning: The Importance of Objectives

Planning starts with an understanding of the objectives of the project. The sponsor, the project manager, and the project team should all have the same understanding about what must be accomplished. There may be several levels of objectives depending on the complexity of the project but all objectives should be interrelated and support the overall goals and acceptance criteria of the project.

The commonly accepted criteria for project objectives are expressed in the word "SMART."

Fill in the blanks with the correct description of each letter of the word SMART as it relates to writing objectives:

S _____

M _____

A _____

R _____

T _____

PMP® &
CAPM®
Exam

Problems with Objectives

Review each of the problems associated with objectives. Explain the error or cause and provide a possible solution or action to prevent the problem from occurring.

Problem	Error/Cause	Recommended Solution
1. Objectives are not agreeable to all stakeholders		
2. Project objectives are too rigid in a project subject to changing priorities		
3. Objectives are not adequately quantified		

Problem	Error/Cause	Recommended Solution
4. Objectives are not documented clearly		
5. Objectives are too complex		
6. Objectives are too numerous		
7. Performance and behavior of the team is less than the established expectation		

Making the grade! An exam tip for you to review.

Project objectives are stated in the Preliminary Project Scope Statement (Integration Management) and further defined and finalized in the Project Scope Statement (Scope Management).

PMP® & CAPM® Exam Tip

6. The Statement of Work (SOW)

PMP® & CAPM® Exam

The statement of work is generally defined as the narrative description of the work to be performed under contract. It provides the potential contractor or bidder with specific information about the work that must be performed and provides a basis for the contractor to decide if their organization can perform the work. In some organizations the statement of work may become the actual signed contract after negotiation. The statement of work should not be confused with the scope, project scope, or project scope statement. (See the glossary for definitions.)

What's wrong with this statement of work?

The contractor agrees to conduct approximately 15 individual tests of the product before it is released to the client for evaluation. The tests will require supervision by the client to ensure that appropriate procedures are utilized during the tests. Test results should be provided to the client shortly after the tests are completed. The client will review the test results and provide feedback before acceptance.

_____

_____

_____

_____

_____

_____

_____

_____

PM Knowledge Note	**Common causes of misinterpretations of the statement of work:**

Common causes of misinterpretations of the statement of work:
Mixing tasks, specifications, approvals, and special instructions
Using imprecise language (nearly, approximately)
No pattern, structure, or chronological order
Wide variation in size of tasks and details
Failing to obtain a "third-party review"

7. Milestone Schedules

In the early stages of project planning before project details are identified, some project scheduling constraints or requirements may be defined in association with contractual agreements. These specific dates or milestones include:

- The project start date
- The project end date
- Other milestones such as phase completions or a requirement for a completion of a deliverable
- Reports or data items
- Certifications and approvals
- A point where future work is restrained

PM Quick Check: A milestone can best be described as:

a) A significant event in the project schedule
b) An activity in the project that consumes a large amount of project resources and time
c) The activities included in the close of a phase or the project

PMP® & CAPM® Exam

8. The Project Charter

The project charter is the document that formally authorizes a project. This means that the project manager, usually assigned at the time of the approval of the charter, is authorized to apply organizational resources to project activities. The charter is initiated and approved, generally, by a project initiator or sponsor external to the organization that will plan and implement the project.

Match the following reasons why a project would be chartered by an organization with the appropriate example:

Reason	Example
_____ **1.** Market demand	a) authorizing a project to meet safety regulations
_____ **2.** Business need	b) a requirement from a client to build a new location to manage expansion
_____ **3.** A customer request	c) a requirement to create a new product to increase revenue
_____ **4.** A technological advance	d) refurbish a public park in the local community
_____ **5.** A legal requirement	e) a new project to produce a faster and more portable handheld computer
_____ **6.** A social need	f) building a fuel-efficient car in response to increasing fuel costs and consumer concerns

What's in a Charter?

List the key items that would generally be included in a project charter.

1. _____

2. _____

3. _____

4. _____

5. _____

6. _____

7. _____

8. _____

9. _____

10. _____

11. _____

PM Knowledge Note

All projects should support the organization's strategic goals and business plans. The strategic goals and business plans should be a major factor in the project selection process.

9. The Project Scope Statement

The project scope statement is a definition of the project and what must be accomplished. It describes in detail the project deliverables and the work required to achieve those deliverables. In some organizations a preliminary scope statement is developed that addresses the characteristics and boundaries or constraints of the project including the deliverables, products, services, and the methods by which the scope will be controlled, verified, and accepted. The scope statement may include the following items:

- Project and product objectives
- Service requirement and product characteristics
- Product acceptance criteria
- Constraints and assumptions
- Schedule milestones
- Initial defined risks
- A high level or order of magnitude cost estimate
- Configuration management requirements
- Approval requirements

The content and detail of the scope statement will vary by organization but the general intent of the scope statement is to provide the project stakeholders, especially the project team, with the information necessary to plan the project in more detail. The project scope statement is developed by answering the basic questions: Who? What? When? Where? Why? How? How much? and How many?

✍ PM Quick Check:

1. What is the difference between a project charter and a project scope statement?

_____

_____

2. What is the difference between "scope" and a project scope statement?

_____

_____

3. Who generally prepares and issues the project charter?

_____

_____

4. What are the main objectives of a project kick-off meeting?

_____

_____

10. Decomposition and the Work Breakdown Structure

The work breakdown structure (WBS) is a deliverable-oriented grouping of project tasks and activities shown in a hierarchal arrangement and in descending levels of detail. The WBS displays the complexity of the project and defines the total scope of the project. The purpose of the WBS is to subdivide the project into smaller, more manageable parts and to provide the appropriate level of detail for further planning. The lowest levels of the WBS are referred to as work packages. The work packages contain the specific activities that must be performed by the project team or assigned resources. These activities are scheduled, their costs estimated, and their progress monitored and controlled.

Making the grade! An exam tip for you to review.

The work package is a deliverable at the lowest level of the WBS where work is to be controlled. The activities in the work package must be performed to produce the deliverable. Work packages are generally associated with a cost account to track the actual costs of the work performed.

PMP® & CAPM® Exam Tip

The process of breaking the project down into smaller, more manageable parts is known as *decomposition*. This process continues until the appropriate level of detail for planning purposes has been achieved. A common practice is to create a WBS template from a similar project to save time and accelerate the planning process. In some cases, where tasks are scheduled to be completed much later in the project, decomposition may not be possible. In these cases a technique referred to as "rolling wave" planning may be used. In this process, the details are added progressively throughout the project, generally phase by phase.

PM Knowledge Note

The work breakdown structure is considered to be the "cornerstone of project planning." It provides the basis for time estimates, cost estimates, risk identification and analysis, and project control. The WBS is also useful for communicating information about the project and building teamwork by including all functional managers associated with the project in the development of the WBS.

WBS Format

The WBS can be developed using several methods. A common approach is to display the WBS in a diagram that resembles an organizational structure. The diagram starts at the highest level, referred to as the project level, and is then broken down into smaller and smaller components and grouped in phases or by deliverables. Another approach is to create an indented format. This format starts with the project name and as each lower level is added, the levels are indented to display the grouping arrangement. Each indentation represents a successively lower level of the WBS. Special note: A WBS is *not* a list of tasks. It is a purposely designed *grouping* of tasks and activities. The lowest level deliverable in a branch of a WBS is referred to as a work package. Within the work package are the activities assigned to resources that will produce the deliverable.

Indentation Example

1.0　Project Name
　　　　1.1 Major project deliverable or subsystem
　　　　　　　　1.1.1 Task 1
　　　　　　　　　　　　1.1.1.1 Subtask 1
　　　　　　　　　　　　1.1.1.2 Subtask 2
　　　　　　　　1.1.2 Task 2
　　　　　　　　　　　　1.1.2.1 Subtask 1
　　　　　　　　　　　　1.1.2.2 Subtask 2
　　　　　　　　　　　　　　　　1.1.2.2.1 Work Package 1

Organization Type WBS

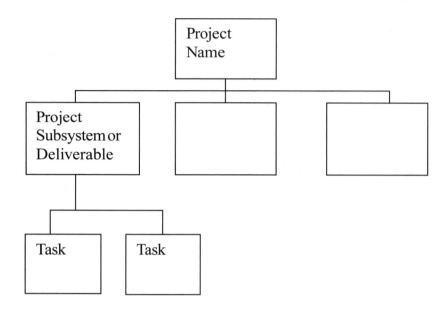

WBS Practice

Using either the indentation technique or the organizational diagram technique, create a WBS for the following project:

Project: Home office addition

Deliverables or Subsystems:

Rough framing and carpentry
Electrical system
Plumbing
Heating and AC
Painting
Decorating and furniture placement

Tasks:

Obtain permits
Purchase material
Frame structure
Run wire
Install pipes
Install heating system
Select paint
Select furniture

11. **Project Selection and Planning: The Role of the Executive**

Before a project is chartered it must be reviewed by the appropriate organization decision makers (executives), who will determine if the project should be undertaken. These selection methods may include an analysis of return on investment, payback period, breakeven analysis, net present value, or internal rate of return. These methods are referred to as *benefit selection methods*. Mathematical models such as decision trees may also be used to determine which projects should be selected. The project executive or executive steering committee generally establishes the selection criteria for projects. Feasibility studies may be required to determine if the project *can* be done and a benefit to cost analysis may be required to determine if the project should be done. In some cases, due to technological issues, competition, compliance requirements, or government regulations, the selection process and criteria may not apply.

a) During project planning and execution the project manager is expected to provide the project executive with information about the project. What information is typically provided to the project executive or sponsor?

_____
_____
_____

b) What is the role of the project executive or sponsor during the project planning process?

_____
_____
_____

✍ PM Quick Check:

True or False

1. The benefits associated with a project may be either tangible or intangible.

_____ True

_____ False

2. Projects are usually prioritized by executives and the connection of the project to the organization's strategic plan.

_____ True

_____ False

3. A WBS displays all project tasks in a logical sequence.

_____ True

_____ False

4. The WBS provides the project manager and team with the ability to develop more reliable cost and schedule estimates.

True _____

False _____

5. The product scope is the work that must be accomplished to deliver the service or result with all specified features and functions.

True _____

False _____

6. The scope statement is the document that provides a common understanding of the scope among all project stakeholders

True _____

False _____

12. Why Projects and Plans Fail

Projects fail for any number of reasons. Many failures can be prevented if the appropriate level of planning is applied. For each of the following reasons of failure, provide a possible solution or preventive measure.

Reason for Failure	Recommended Action or Solution
1. Corporate goals are not understood at the lower or organizational levels or functional units	
2. Plans are too complex and require too much in too little time	
3. Financial estimates are significantly inaccurate	
4. Plans are based on insufficient data	
5. Planning is performed by a planning group	
6. No one knows the ultimate objective of the project	
7. Staffing requirements are not clearly understood	
8. The project personnel lack the required skills	
9. The project team is not working toward the same specifications	
10. The project team is constantly being shuffled in and out of the project	

13. Management Control and Configuration Control

After the project has been planned and the plans have been approved by the key stakeholders, a monitoring and control process should be utilized to prevent unauthorized work from being performed, scope creep from developing, and variances to the baselines from exceeding established thresholds. Project control is concerned with:

- Comparing actual project performance to the planned baselines
- Assessing performance to determine if corrective action is necessary
- Identifying, tracking, and responding to risks
- Providing status to key stakeholders
- Providing updated cost and schedule forecasts
- Managing changes through a specified process

A project methodology and a system for storing project information (Project Management Information System or PMIS) provide the project manager and team with useful information to assist in maintaining control and providing information to stakeholders. Upper management or the project sponsor may establish guidelines regarding control to meet organizational requirements or conform with established policies.

Configuration management is a form of change control. It is a process or technique that manages changes to the features, functions, or physical characteristics of the deliverables of the project or the final product. A configuration control board (CCB) or committee reviews requested changes to determine if a change should be approved.

Managing Configuration Changes

If you were a member of a configuration control board, what questions would you ask before determining whether a change should be approved or not?

_____

_____

_____

_____

_____

_____

14. Project Phase-Outs, Transfer, and Closing

Projects have a definite end date (at least by definition), and therefore, transfer of project deliverables to the customer or intended user and close out of the project should be planned. The closeout of the project is actually a project in itself and a very important phase of the project. Projects are generally closed based on the following reasons:

Completion of all project objectives

Achieving the desired value

- Achieving agreed upon contractual terms and conditions
- Preparing for the transition to the operational phase of the project's product
- Analysis and documentation of overall project performance
- Identification of possible follow-on business

The project is no longer associated with organizational goals, will not achieve its targeted value, or has reached a point where it is no longer perceived as financially beneficial.

What activities should be included in the close out or transfer of the project deliverable to the operational environment? Make a list of activities that should be included in the close out phase.

1. _____

2. _____

3. _____

4. _____

5. _____

6. _____

7. _____

8. _____

Kerzner "Quick tips" for the Project Management Institute PMP® and CAPM® EXAM

The topic "Planning" is related to all of the knowledge areas of the *PMBOK® Guide*. Planning involves the integration of all project knowledge areas. Specific references about this chapter are mainly located in Project Integration Management and Project Scope Management.

The PMP® Exam will include many questions that are arranged in the form of scenarios. A suggested study strategy is to apply the principles, tools, and techniques described in this chapter to actual projects. Experience in using the tools and techniques will assist in improving your understanding about a concept and how the tools and techniques can assist in achieving desired results.

Remember the differences between the definitions of project scope statement, statement of work, and scope of the project.

The charter authorizes the existence of the project and usually names/assigns the project manager. The charter is not a plan or the scope statement. It is an input to scope planning.

The WBS is a key part of total project planning and provides a basis for estimating costs, schedules, resources, and for identifying project risks. It is not a list; it is a grouping of tasks and activities and facilitates project planning. The WBS is often referred to as the "cornerstone of project planning."

Configuration Control   Also known as change control, this process is used to effectively manage, review, and approve recommended or requested changes to the products and deliverables of the project. Configuration management is generally associated with the management of the features, functions, and physical characteristics of products.

Scope Planning   This process defines the objectives (objectives should be SMART) and who, what, when, where, why, how, and how much. The output is the scope management plan.

Scope Definition   This process uses the project charter, preliminary scope statement, product analysis, and stakeholder analysis to develop the project scope statement.

Project Scope Management   Scope management in the 4th edition of the *PMBOK® Guide* includes the following processes: Collect requirements, Define Scope, Create WBS, Verify Scope, and Control Scope

---

Additional tips and practice items for the PMP® exam are included in each chapter of the 10th edition and in the section of the workbook entitled *PMP® Exam and PMBOK Guide® Review.*

Answers to Questions and Exercises

1. The Logic of Planning

   1. a   **5.** b
   2. e   **6.** h
   3. g   **7.** d
   4. f   **8.** c

2. PM Quick Check

   1. True
   2. False
   3. True

3. The benefits of developing and using a standard project life cycle are: increased manageability, standardization, and more effective control.

PM Quick Check—Question 1 – B Question 2 - All answers are correct

4.

Responsibility	Project Manager	Functional Manager
Set project goals and objectives	X	
Develop detailed task descriptions		X
Communicate/Establish major milestones	X	
Set high-level requirements	X	
Provide detailed schedules		X
Identify areas of risk, uncertainty, and conflict		X
Establish ground rules, assumptions	X	
Provide summary status reports for the sponsor	X	

5. Specific, measurable, attainable or action oriented, realistic, time based

Problem	Error/Cause	Recommended Solution
1. Objectives are not agreeable to all stakeholders	Failure to obtain input from the stakeholders	Identify all stakeholders and obtain requirements
2. Project objectives are too rigid in a project subject to changing priorities	Lack of flexibility in planning to accommodate potential changes and uncertainty	Set objectives based on the "SMART" concept
3. Objectives are not adequately quantified	No method of measurement	Establish specific metrics for each objective
4. Objectives are not documented clearly	Failure to obtain feedback from stakeholders and the project team	Communicate, obtain feedback, and refine objectives to ensure full understanding
5. Objectives are too complex	Failure to define objectives at the appropriate level of detail and establish realistic expectations	Review objectives and break down into more manageable elements
6. Objectives are too numerous	Too many objectives can cause concerns about span of control and could overwhelm the project team or cause mismanagement of resources	Reduce objectives to an acceptable number (5–10) Break higher-level objectives into smaller elements and assign to functional managers or units
7. Performance and behavior of the team is less than the established expectation	Failure to establish expectations early in the project. Unclear objectives. Poor communication to the team. No performance appraisal process	Establish expectations at the project kickoff and ensure that all objectives are communicated clearly; achieve buy-in. Establish a process for regular performance reviews

6. What's wrong with this statement of work? Use of the word "approximately," no description of the term "supervision," no explanation of "appropriate procedures." Use of the word "shortly." No explanation of how the client will review the test results or acceptance criteria.

7. PM Quick Check—a

8.
    1. f
    2. c
    3. b
    4. e
    5. a
    6. d

What's in a charter?

1. Business need
2. Project purpose or justification
3. The assigned project manager
4. Summary milestones
5. Stakeholder influences—who will be affected
6. Assumptions
7. Functional groups that will be involved
8. High-level or summary budget
9. Sponsor name

9. PM Quick Check:

   1. A charter authorizes the project. The scope statement defines the project in detail and includes the specific objectives.
   2. Scope is the sum of all products and services to be delivered as a project. The project scope statement is a narrative description of the project scope and includes the major deliverables, objectives, assumptions, and constraints of the project.
   3. The sponsor or project initiator.
   4. Introduce the project manager to the team, assemble the subject matter experts, explain project success criteria, communicate project objectives, define team roles and responsibilities, explain constraints and assumptions, begin the detailed planning process

10. WBS example

    1.0 Home Office Addition
        1.1 Rough Framing and Carpentry
            1.1.1 Obtain permits
            1.1.2 Purchase material
                1.1.2.1 Determine lumber requirements
        1.2 Electrical System
        1.3 Plumbing
        1.4 Heating and AC

11. a) Project variances in cost and schedule, achievement of milestones, major delays or changes in project plans, risk situations, customer satisfaction issues,

    b) Select the project manager, define high level project requirements, provide resources, communicate organizational strategies, liaison to other business units or functional groups.

    1. True
    2. True
    3. False
    4. True
    5. False – The project scope is the work that is required to produce the product or service.
    6. True

12. Why Projects and Plans Fail

Reason for Failure	Recommended Action or Solution
1. Corporate goals are not understood at the lower or organizational levels or functional units	Review corporate goals with the sponsor and then communicate to the project team. Provide an opportunity for feedback to ensure understanding
2. Plans are too complex and require too much in too little time	Review the project objectives and consider breaking into smaller separate projects introduced in phases
3. Financial estimates are significantly inaccurate	Establish reliable and consistent estimating practices. Use lessons learned and previous projects for comparison
4. Plans are based on insufficient data	Conduct feasibility studies and cost benefit analysis. Conduct risk assessment. Obtain expert knowledge
5. Planning is performed by a planning group	Arrange to have plans developed by the functional units that will perform the work
6. No one knows the ultimate objective of the project	Establish clear roles and responsibilities at project kickoff. Prepare and communicate a project scope statement
7. Staffing requirements are not clearly understood	Create a WBS, use a scope statement and the project charter to determine staffing and resource needs
8. The project personnel lack the required skills	Establish a screening process for staffing or schedule training for team members
9. The project team is not working toward the same specifications	Define project specifications and requirements early in the project. Document and distribute specifications. Conduct reviews
10. The project team is constantly being shuffled in and out of the project	Establish formal signed agreements from functional units when obtaining project staff. Arrange for contingencies or back-up personnel. Cross-train. Escalate to the project sponsor

13. Managing Configuration Changes

Why is the change necessary? How much will it cost? Who will do it? How will the change affect the schedule, scope, or budget? What are the risks?

14. Close Out Phase Activities

1. Prepare all documents and project records
2. Conduct post-project reviews
3. Obtain acceptance from the client and the sponsor
4. Verify compliance with contract requirements
5. Release resources
6. Prepare performance appraisals
7. Close out all open work orders
8. Prepare final project financial reports and close financial accounts

Your Personal Learning Library

Write down your thoughts, ideas, and observations about the material in the chapter that may assist you with your learning experience. Create action items and additional study plans to assist you in enhancing your skills or for preparing to take the PMP® or CAPM® exam.

Insights, key learning points, personal recommendations for additional study, areas for review, application to your work environment, items for further discussion with associates.

_____

_____

_____

_____

_____

_____

_____

_____

_____

_____

_____

Personal Action Items:	
Action Item	**Target Date for Completion**

Chapter Twelve

NETWORK SCHEDULING TECHNIQUES

The project scope statement and Work Breakdown Structure provide the project manager and team with the foundation for further planning. The scope statement and WBS answer the what, when, where, why type of questions but additional planning is required. The tasks and activities of the project must be arranged in the appropriate sequence to meet schedule requirements, mandatory dependencies, resource constraints, and other factors that affect the total or integrated project plan. Network scheduling is actually a continuation of the project planning process and allows the project team to determine the best approach to managing and controlling project activities. The process of network scheduling provides an opportunity to display the interdependencies of project activities; where float or slack exists, where there is flexibility in when activities may be performed, the critical path and opportunities for rearranging activities ensure the most efficient use of resources and available time.

**PM
Knowledge
Note**

The WBS is generally considered to be the "cornerstone of project planning" and provides the basis for network scheduling and the development of network diagrams. The project team collaborates to identify the project components, major tasks and work package activities and develops the WBS to clearly display the project and its complexity.. Network scheduling generally follows the development of the WBS. The activities defined in the WBS are placed in a logical sequence using the input of the project team to display the dependencies and relationships of each of the activities of the project.

The major advantage of network scheduling is the ability to display interdependencies. Gantt charts, milestone charts, and the WBS are useful for planning and control but do not effectively allow the project team to position activities in a logical sequence. The analysis associated with network scheduling also provides valuable information for total integrated planning, time studies, improved schedule development, resource management, and re-planning.

Glossary of terms   Key terms and definitions to review and remember

Activity   An element of work performed during the course of a project. Each project activity generally has a defined duration, an estimated cost, and specific resource requirements.

Activity-on-Node   Commonly referred to as Precedence Diagram Method. The activity is depicted as the node or box. Each activity in the diagram is connected using arrows to indicate dependencies and logic relationships.

Arrow Diagram Method (ADM)   A network diagramming technique in which activities are represented by arrows. Activities are connected at points called nodes to illustrate the sequence in which activities are expected to be performed. This is also known as Activity-on-Arrow (AOA).

Backward Pass   The calculation of late finish and late start dates for project activities when using the critical path method.

Baseline   The original approved plan. The three main project baselines are scope, schedule, and budget. The baseline may be updated periodically to include approved changes.

Crashing   A technique of schedule compression where the total project duration is reduced by identifying activities that can be reduced in duration by adding additional resources and cost. Crashing focuses on seeking opportunities for compressing the project schedule by targeting activities that will provide the greatest duration reduction at the lowest cost. Crashing generally is utilized on the critical path where there is no slack or float.

Critical Activity   An activity on a critical path.

Critical Path   The longest path through a network that determines the earliest completion date of the project. The critical path is usually defined as the path through a network where the activities have zero float or slack.

Critical Path Method (CPM)   A network analysis technique used to determine the early start, early finish, late start, and late finish dates for activities on the various logical paths of a network. Early dates are calculated using a forward pass; late dates are calculated using a backward pass.

Dummy Activity   An activity of zero duration used to show a logical relationship. .Dummy activities are used in the arrow diagramming method (activity on the arrow) when special logical relationships are required and must be displayed Dummy activities are shown graphically as a dashed or dotted line headed by an arrow and connecting two nodes. Dummy activities may be a factor in calculating the critical path.

Duration   The number of work periods required to complete an activity or other project element. Duration is generally estimated by the functional manager responsible for the activity.

Effort   The number of labor units required to complete an activity or other project element. Effort is associated with skill level, resource competency, and number of resources available. Effort is related to the determination of activity duration.

Fast Tracking   Compressing the project schedule by overlapping activities that would normally be done in sequence. Risk must be considered when utilizing fast tracking.

Float   The amount of time that an activity may be delayed from its early start without delaying the project finish date. (Also referred to as slack.)

Forward Pass   The calculation of the early start and early finish dates for activities in a network diagram.

Free Float   The amount of time an activity can be delayed without delaying the early start of any immediately succeeding activities.

Hammock   A group of related activities aggregated to a summary level. Also known as a summary activity.

Lag   A modification of a logical relationship that directs a delay in the successor task. Example: After concrete is poured (the activity is complete), there is a waiting period to allow the concrete to cure. This may take several days. The next succeeding activity cannot start until the curing time is satisfied. The curing time or elapsed time is known as lag.

Lead   A modification of a logical relationship that allows an acceleration of the successor task. Example: Task B depends on the completion of task A. The team determines that task B can start two days before the end of task A. The lead for task B is two days.

Level of Effort (LOE)   A support type activity that does not readily lend itself to measurement of discrete accomplishment. Level of effort is considered when determining the duration of an activity. Example: The level of effort for an expert to perform an activity would be much less than the level of effort of a newly trained employee. A project manager provides a supporting effort to achieve project objectives.

Master Schedule   A summary level schedule that identifies the major activities and milestones associated with a project.

Milestone   A significant event in the project. Examples: end of a phase, end of the project.

Precedence Diagram Method (PDM)   A network diagramming technique in which activities are represented by nodes. Activities are linked by precedence relationships to show the sequence in which the activities are to be performed. There are four logical relationships in the PDM: Finish to start, finish to finish, start to start, start to finish.

Program Evaluation and Review Technique (PERT)   An event-oriented network analysis technique used to estimate project duration when there is a high degree of uncertainty with the individual activity duration estimates. Associated with the formula: The sum of the optimistic time plus four times the most likely time, plus the pessimistic time divided by six. This is the PERT weighted average formula.

Project Network Diagram   Any schematic display of the logical relationships of project activities.

Resource Breakdown Structure   Hierarchal structure of the identified resources by resource category and resource type.

Rolling Wave Planning   Adding more details as the project progresses to ensure the most current information known about the project. This practice is called rolling wave planning because the planning wave stays ahead of the work being executed.

Slack   Generally considered to have the same meaning as float.

Total Float   The amount of time that a scheduled activity may be delayed from its early start date without delaying the project finish date.

Time Management Processes—A Brief Review

The following processes are generally associated with the knowledge area Project Time Management. Network scheduling is directly associated with the Time Management process of activity sequencing and is an essential element of the integrated planning process.

The *PMBOK® Guide*—Fourth edition includes the following Project Time Management processes:

Define Activities – further development of the WBS at the work package level to identify specific actions that must be taken to complete project deliverables

Sequence Activities – identifying the specific logical relationships of project activities

Estimate Activity Resources – determining the appropriate resource types and quantities required to complete project activities

Estimate Activity Duration – determining the number of work periods required to complete project activities

Develop Schedule – analysis of project resources, activity sequences, activity durations, and project constraints to determine the project schedule

Control Schedule – monitoring the project schedule to identify variances, managing project changes and updating the project schedule as required

Define activities

- The process of identifying and documenting the specific activities that must be performed to produce the various project deliverables identified in the WBS. This is a continuation of the process of decomposition. The activities are defined to the detail that will allow project objectives to be met.

- **Input includes:** Enterprise environmental factors, organizational process assets, scope baseline. Other inputs may include: WBS, project scope statement, WBS dictionary, historical information, constraints, assumptions, and the project plan.

- **Tools and techniques:**

  - Decomposition: Involves subdividing project elements into smaller, more manageable components.
  - Templates: Activity lists or portions of activity lists from previous projects.
  - Expert judgment.
  - Rolling Wave planning - continuous process of planning near term tasks in detail and maintaining some focus on future tasks

- **Outputs:** An activity list, milestone list, activity attributes. Other outputs may include: requested changes (to the WBS and scope statement), supporting details including updated assumptions and constraints, and updates to the WBS.

Sequence Activities

- The process of identifying and documenting activity dependencies and relationships.

  - There are three common types of dependencies:
    - Mandatory—also known as hard logic. Tasks must be completed in a specific sequence due to physical necessity.

- Discretionary—also known as soft logic (discretionary dependencies are associated with best practices)
- External—dependency associated with another project or other item outside of the project and generally affecting the critical path.

- **Inputs:** Activity list, activity attributes, milestone list, project scope statement, organizational process assets. Other inputs may include: approved changes, product description, identified dependencies, constraints, and assumptions.

- **Tools and techniques:** Precedence Diagramming Method (Activity-on-node, dependency determination, applying leads and lags, schedule network templates. Other techniques include: Arrow Diagramming Method (Activity-on-Arrow), conditional diagramming methods such as GERT – Graphical Evaluation Review Technique.

- **Outputs:** Project schedule network diagram, project document updates. Other outputs may include: activity list updates, activity attribute updates, and requested changes (the sequencing activity may generate a need to change an activity or refine the schedule).

Estimate Activity Resources

- The process of estimating the number of resources required to complete each identified activity.

- **Input includes:** Enterprise environment factors, organizational process assets, activity list, activity attributes, resource calendars, project management plan.

- **Tools and techniques:** Expert judgment, alternatives analysis, published estimating data, project management software, bottom-up estimating (performed using the WBS as a reference).

- **Outputs:** Activity resource requirements, activity attribute updates, resource breakdown structure, resource calendar updates, requested changes.

- Resources that are associated with the estimating process include:
  - Equipment
  - Materials
  - Supplies
  - Money
  - People (skill levels, competency, availability)

Estimate Activity Duration

- The process of estimating schedule activity durations using information from the scope statement, identified resource types, number of resources available, calendars, ramp-up time (the affect of non-contiguous work assignments).

- **Input includes:** Activity list, activity attributes, enterprise environmental factors, organizational process assets, project scope statement, activity resource requirements, resource calendar, project scope statement. Other inputs may include: the project management plan including risk register, and activity cost estimates.

- **Tools and techniques:** Expert judgment, analogous estimating, parametric estimating, three-point estimates, reserve analysis.

- Three-point estimates are based on determining the optimistic, most likely, and pessimistic durations of an activity. The weighted average estimate or estimated time is calculated by the formula: Sum of the optimistic estimate plus four times the most likely estimate plus the pessimistic estimate divided by six.
- Analogous estimating uses actual duration figures from similar activities or projects. Also referred to as a top-down type of estimating. This type of estimating is generally used early in a project. Estimates are refined as more project data is obtained.
- Parametric estimating calculates duration estimates by using mathematical algorithms such as multiplying the quantity of work in work periods by the cost rate of a resource. This type of estimate works best for standardized and often repetitive activities. This is also a top-down type of estimating. Example: Multiplying the cost per square foot by the desired number of square feet of living space will produce a parametric estimate of the cost of building a home.

- **Outputs:** Activity duration estimates, project document updates

Develop Schedule

- An iterative process to determine planned start and end dates for project activities. Schedule development involves the combined input of the project team and includes negotiation, agreement about contingencies, availability of critical resources, and contractual agreements.

- **Input includes:** Organizational process assets, enterprise environmental factors project scope statement, activity list, activity attributes, project schedule network diagram, activity resource requirements, resource calendars, Activity Duration Estimates,

- **Tools and Techniques:** Schedule network analysis, critical path method, schedule compression including crashing and fast tracking, "what if" scenarios, resource leveling, critical chain method, project management software, applying calendars, adjusting leads and lags, scheduling tools such as: schedule model (use of a network diagram or other model for analysis).

- **Outputs:** Project schedule and supporting detail, project schedule network diagram, schedule baseline, project calendar, project document updates. Other outputs may include: activity attributes updates, Gantt charts or bar charts, milestone charts, requested project plan updates.

Control Schedule

This process is used by the project team to determine the status of the project schedule, identifies factors that can change or impact the project schedule, provides a process to determine if variances to the schedule or a change in the schedule has occurred, and manages actual changes upon approval. Schedule control is an element of Integrated Change Control.

- **Inputs Include:** Project management plan, schedule baseline, work performance information. Other inputs may include: reports from Earned Value Management or Variance Analysis and approved change requests.

- **Tools and Techniques:** Performance reviews, variance analysis, resource leveling, what if scenarios, adjusting leads and lags, schedule compression, scheduling tools. Other tools may include: Progress reports, schedule change control system, performance measurement, project management software.

- **Outputs:** Work performance measurements, organizational process assets updates, change requests, project management plan updates, project document updates. Other outputs may include: Schedule baseline updates, recommended corrective action, activity list updates.

Activities, Questions, and Exercises

Refer to Chapter Twelve of *Project Management: A Systems Approach to Planning, Scheduling, and Controlling* (10th Edition) for supporting information and assistance in completing each exercise.

The following questions and exercises are most closely associated with the knowledge areas of the *PMBOK® Guide*: Project Time Management. It is important to remember that although each knowledge area is presented separately in the *PMBOK® Guide* for learning purposes, many processes are performed concurrently and all processes are associated with Project Integration Management.

Review each of the following questions or exercises and provide the answers in the space provided. Some questions and situations will have more than one appropriate answer or response. Project managers are aware that many issues they face on a daily basis may have several possible solutions. These exercises are designed to encourage you to think about alternatives before taking action.

Activity Planning—From WBS to Project Network Diagram

After the project has been defined by developing a work breakdown structure, the next step in the planning process is to perform the steps required to develop the project schedule. This process includes identifying activities, determining resource requirements, estimating activity durations, sequencing, and managing within project constraints.

1. ✍ **PM Quick Check:** Defining Activities

   Complete the following statements:

   A. The process of activity definition is an extension of the development of the _____.

   B. After the major project deliverables have been defined, the lower level project deliverables and

   activities are included in the _____.

2. Activity Definition—Inputs, Tools and Techniques, and Outputs

   Review the list of items and determine if each item is an input, a tool or technique, or an output. Place each item in the correct column:

PMP® & CAPM® Exam Tip

Inputs	Tools and Techniques	Outputs

a) Enterprise environmental factors
b) Decomposition
c) Templates
d) Activity list
e) Organizational process assets
f) Rolling wave planning
g) Project scope statement
h) Expert judgment
i) WBS dictionary
j) Planning component
k) Requested changes
l) Activity attributes
m) Milestone list
n) Project management plan
o) WBS

3. Sequencing Activities and Network Diagrams

The next process following activity definition involves arranging the activities from the activity list into a specific sequence based on dependencies identified by the project team. Some activities depend on input from another activity before they can begin or are constrained by either time, resources, or other factors. The sequencing process identifies all relationships between activities and ensures that there are no "hanging activities." A hanging activity is one that does not have clearly defined relationships and is not completely connected to the network.

A. Precedence Diagramming Method—Study Note

The Precedence Diagramming Method is used to display project activities using nodes that are linked together using one or more types of logical relationships. The logical relationships are shown using arrows to connect the nodes. PDM uses four types of logical relationships—finish to start (the most common relationship), finish to finish, start to start, and start to finish.

3-a Exercise: Create a network diagram using the PDM and the information in the table: (Hint: A common practice in the development of network diagrams is to use Post–It ™ notes or other similar products that allow the team to move activities around until they are satisfied that the diagram correctly displays the project network. Establishing a start and an end milestone will assist in making sure there are no "hangers.")

Activity	Predecessor
Start	none
A	Start
B	A
C	A
D	B
E	C
F	D
G	E,F
End	G

B. Arrow Diagramming Method—Study Note

The Arrow Diagramming Method (ADM) is similar to the PDM, except that when using ADM, all dependencies are shown as finish to start. The activity is on the arrow instead of the node and durations are shown on the arrows. ADM is also known as Activity-on-Arrow (AOA). A dummy activity may be used in the ADM to show a logic relationship. Dummy activities do not show duration. They are used to indicate a relationship.

3-b Exercise: Create a network diagram using the Arrow Diagramming Method and the information in the table.

Activity	Name	Predecessor/Dependency
1-2	A	None
2-3	B	A
2-4	C	A
3-5	D	B
3-7	E	B
4-5	F	C
4-8	G	C
5-6	H	D,F
6-7	I	H
7-8	J	E,I
8-9	K	G,J

PMP® Exam and CAPM® Exam study notes:

In normal distribution, measurements are assessed by the number of standard deviations they appear from the mean. In normal distribution the following information is important to remember:

- 68.3% of the data points fall within one standard deviation from the mean

- 95.5% of the data points fall with two standard deviations from the mean

- 99.7% of the data points fall within three standard deviations from the mean

Basic formula for standard deviation – (pessimistic value minus the optimistic value) divided by 6.

Note: This is a short formula generally accepted for calculating standard deviation.

Mean - general the average of all data points

Median – the midpoint of the distribution where 50% of the values are above and 50% of the values are below the data point. Example: In the series 1,2,3,4,5,6,7, the median is 4

Mode – the most frequent data point value. Example – 1,1,2,2,2,2,3,4,4,5,6, the mode is the number 2

Range – the difference between the highest value data point and the lowest value data point. Example: 10, 20, 40, 50, 70. 80, 100 the ranges = 100 - 10 or 90

4.   ✎ **PM Quick Check:**

A)   List the four types of logical relationships that may be utilized in the Precedence Diagram Method.

   1. _____

   2. _____

   3. _____

   4. _____

B)   The Arrow Diagramming Method uses only a (1) _____ relationship and may

   use (2) _____ that do not have a duration to complete the logic of the diagram.

   1. Finish to Finish (FF) or Finish to Start (FS) or Start to Start
   2. Nodes or Dummy Activities or Tasks Summary Activity

C)   The most common relationship in a network using PDM is

   _____

PMP® &
CAPM®
Exam

Study Notes

Activity duration estimating is used to determine the number of work periods required to complete a schedule activity. Duration is determined by analyzing several factors including the number of resources required, the competency of the resources, and the availability of the resources. Work periods may be represented in several ways including days, weeks, or hours based on the preference of the project team and stakeholders. The process of activity duration estimate may follow activity definition, but in many cases the estimates are provided concurrently.

A PERT Chart is a type of network diagram that uses probabilistic estimates to determine the project duration. PERT is the abbreviation for Program Evaluation Review Technique. Network diagrams are commonly referred to as PERT Charts, but a PERT Chart is actually a network diagram that is constructed using the critical path method (CPM) and a weighted average formula. The formula is expressed as

$$\frac{a + 4m + b}{6}$$

where: a = optimistic, b = pessimistic, and m = most likely. The standard deviation may be calculated using the formula:

$$\frac{b - a}{6}$$

This formula will provide an acceptable value for the standard deviation. Calculating the standard deviation provides a basis for establishing a confidence level for each activity on the critical path. Using a normal distribution, it is known that there is a 68% probability of completing the activity on the project within one standard deviation from the mean, 95% within two standard deviations, and 99.73% within three standard deviations.

A GERT Chart or Graphic Evaluation Review Technique is a diagram that allows looping, branching, and multiple end points. A PERT Chart does not allow feedback loops and conditional branching.

Types of dependencies. There are three basic types of dependencies:

- Mandatory dependencies or hard logic. This type of dependency is inherent in the nature of the work. They often involve physical limitations. The work must be performed in a specific order. Example: Install walls before adding the roof.
- Discretionary or soft logic. Generally, the dependency is determined by the project manager or team and is associated with best practices. "It is best to do it this way."
- External dependencies are those that are associated with factors that are outside of the project but may impact the critical path.

5. Forward Pass

The forward pass is used to determine the earliest start and earliest finish dates of each activity on the network diagram. The process begins with the first activity on the network and continues from left to right. During the process, at any point where two or more activities converge on another activity the higher value of the early finish is used to determine the early start of the succeeding activity. This process continues through each path in the network to determine the total duration of the project and the critical path.

Exercise: Using the information provided in the following table, create a network diagram using the precedence diagram method and determine the duration of the project and the critical path.

Activity	Predecessor	Duration
Start		
A	None	5
B	A	2
C	A	3
D	B	7

Activity	Predecessor	Duration
E	C	4
F	D	1
G	E,F	2
End	G	

Project Duration _____

Critical Path _____

6. Backward Pass

The backward pass is used to determine the latest start and latest finish of each activity. The difference between the latest finish and the earliest finish will determine the slack for the activity. There is generally no slack on the critical path. The backward pass is performed on the network diagram from the end or right and back through the network to the starting point. At points of convergence the smaller number is used to determine the latest finish of a preceding task.

Referring to the diagram developed in exercise 6, perform a backward pass to determine the late start and late finish of each activity.

A. Identify the activities that have slack.

Activity	Predecessor	Duration	Slack
Start			
A	None	5	
B	A	2	
C	A	3	
D	B	7	
E	C	4	
F	D	1	
G	E,F	2	
End	G		

B. What is the early start of Activity C? _____

C. What is the late start of Activity E? _____

7. ✍ **PM Quick Check:**

1. Activities with zero time duration are referred to as:

   a) critical activities
   b) noncritical activities
   c) events
   d) dummy activities

2. In the Precedence Diagramming Method, the most common logical relationship is

   a) start to finish
   b) finish to start
   c) finish to finish
   d) start to start

PMP® &
CAPM®
Exam

8. **Crashing the Schedule**

During the project scheduling process or during actual execution of the project, it may become necessary to shorten the duration of the project. Crashing generally means to reduce duration by using additional resources or overtime. Review the information in the table. Determine which activities should be crashed. (Hint: Determine the critical path first.)

Activity	Predecessor	Normal time required in weeks	Crash time in weeks	Normal cost	Crash cost	Crash? Yes/No
A	None	4	2	10,000	14,000	
B	A	6	5	30,000	42,500	
C	A	2	1	8,000	9,500	
D	B	2	1	12,000	18,000	
E	B,C	7	5	40,000	52,000	
F	D,E	8	3	20,000	29,000	

9. **Lead and Lag**

During project planning it may be necessary to adjust the project schedule to include lag time, a delay of the succeeding activity or lead time, an acceleration of a succeeding activity. A common example of lag time is the elapsed time after the activity of pouring concrete. The concrete must cure for a period of time. Lag is built into the network to allow for the curing time. An example of lead time is to indicate that a succeeding activity may start after 70% of a preceding activity has been completed.

In the following diagram, if the early start of activity A is 10 days, what is the early finish time for activity C?

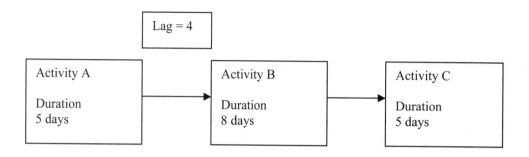

10. **Network Scheduling Practice**

A. Using the PDM, draw the network diagram, identify the critical path, and calculate the early start and early finish times for each activity.

Activity	Dependency	Activity duration (weeks)	Early start	Early finish
A	None	4		
B	None	6		
C	A,B	7		
D	B	8		
E	B	5		
F	C	5		
G	D	7		
H	D,E	8		
I	F,G,H	4		

B. Fun with PERT. Draw an arrow diagram for the following project. What is the critical path? What is the duration of the project? What is the standard deviation for Activity G? (Te = estimated time)

Activity	Initial node	Final node	Optimistic time	Pessimistic time	Most likely time	Te
A	1	2	1	3	2	
B	1	4	4	6	5	
C	1	3	4	6	5	

Activity	Initial node	Final node	Optimistic time	Pessimistic time	Most likely time	Te
D	2	6	2	4	3	
E	2	4	1	3	2	
F	3	4	2	4	3	
G	3	5	8	15	10	
H	4	6	4	6	5	
I	4	7	6	14	10	
J	4	5	1	3	2	
K	5	7	2	4	3	
L	6	7	6	14	10	

11. ✐ **PM Quick Check:** The critical path is the

  a) Shortest path through the network diagram.
  b) Longest path through the project network that determines the earliest time the project can be completed.
  c) Path with the greatest number of resources assigned.

12. You have received the following information from a project team member: the estimated pessimistic duration to complete an activity is 45 days, the optimistic time is 15 days and the most likely time is 25 days. What will you post as the duration for this activity using normal distribution?

_____

_____

13. Using the same information from the previous question calculate the standard deviation using the short formula

_____

14. Using normal distribution, what is the probability of completing the activity referred to in questions 12 and 13 within 21.67 days and 31.67 days?

_____

15. What is the probability of completing the activity in 31.67 days or less? _____

Making the grade! An exam tip for you to review.

The PMP® Exam will test your knowledge of the forward pass, backward pass process. Practice the critical path method often and remember the rules: use the greater number when performing the forward pass at points of convergence and the smaller number at points of convergence in the backward pass process.
Remember the PERT weighted average formula and the formula for standard deviation.

PMP® & CAPM® Exam Tip

Kerzner "Quick tips" for the Project Management Institute PMP® and CAPM® EXAM

The topic "Network Scheduling Techniques" is most closely related to the following areas in the *PMBOK® Guide*: PROJECT TIME MANAGEMENT

Review the inputs, tools and techniques, and outputs of the time management processes: Activity Definition, Activity Sequencing, Activity Resource Estimating, Activity Duration Estimating, Schedule Development, and Schedule Control.

The PMP® exam will test your ability to recognize and understand two types of network diagrams: the *precedence diagramming method (PDM)* and the *arrow diagramming method (ADM)*. Make sure you know the differences between them. PDM has four possible logical relationships:

- **Finish-to-start (the most common dependency type)**—The successor activity's start depends on the completion of the successor activity.

- **Finish-to-finish**—The completion of the successor activity depends on the completion of the predecessor activity.

- **Start-to-start**—The start of the successor activity depends on the start of the predecessor activity.

- **Start-to-finish**—The completion of the successor activity depends on the start of the predecessor activity.

ADM uses only the finish-to-start relationship and may use dummy activities to complete the logic relationships.

Become familiar with different types of scheduling charts: Gantt, milestone, and schedule networks.

Be prepared to explain the critical path method and how to determine critical path.

Understand the difference between the schedule compression techniques of crashing, fast tracking, and managing slack.

Understand how to calculate the activity durations using PERT.

Practice forward pass and backward pass calculations to determine where float or slack exist.

Know how to calculate early start, early finish, late start, and late finish.

Additional tips and practice items for the PMP® exam are included in each chapter and in the section of the workbook entitled *PMP® Exam and PMBOK® Guide Review.*

Answers to Questions and Exercises

1. **A** WBS   **B** Work Package

2.

Inputs	Tools and Techniques	Outputs
a	b	d
e	c	k
g	f	l
n	h	m
i	j	
o		

3. **A**
   B

4. **A**—Finish to start, finish to finish, start to start, start to finish
   B—**1.** finish to start; **2.** dummy activities
   C—finish to start

5. Project Duration = 17   Critical Path = ABDFG

6. Backward Pass

   A C and E have slack (3 units each)
   B 5
   C 11

7. 1—d, 2—b

8. Tasks A,B,E,F could be considered for crashing. Always look for the least cost to achieve the desired crashing effect and watch out for new critical paths that may appear.

9. 32 days

10. A.

Activity	Dependency	Activity duration (weeks)	Early start	Early finish
A	None	4	0	4
B	None	6	0	6
C	A,B	7	6	13
D	B	8	6	14
E	B	5	6	11
F	C	5	13	18
G	D	7	14	21
H	D,E	8	14	22
I	F,G,H	4	22	27

B.

Activity	Initial node	Final Node	Optimistic time	Pessimistic time	Most likely time	Te
A	1	2	1	3	2	2
B	1	4	4	6	5	5
C	1	3	4	6	5	5
D	2	6	2	4	3	3
E	2	4	1	3	2	2
F	3	4	2	4	3	3
G	3	5	8	15	10	10.5
H	4	6	4	6	5	5
I	4	7	6	14	10	10
J	4	5	1	3	2	2
K	5	7	2	4	3	3
L	6	7	6	14	10	10

Critical Path is CFHL.   Duration is 23.   Standard deviation of G = 1.17.

11. b

12. 26.67 days

13. The standard deviation is 5 (45 – 15 / 6)

14. 68.3%   The probability associated with 1 standard deviation in normal distribution is 68.3%. in this example one standard deviation below the mean is 26.67 – 5 = 21.67. 26.67 + 5 = 31.67

15. The probability of finishing the activity in 31.67 days or less is 84%. The mean is 50% plus 68% /2 = 34%.. 31.67 is one standard deviation from the mean. Add the mean plus ½ of the 68% = 84%.

Your Personal Learning Library

Write down your thoughts, ideas, and observations about the material in the chapter that may assist you with your learning experience. Create action items and additional study plans to assist you in enhancing your skills or for preparing to take the PMP® or CAPM® exam.

Insights, key learning points, personal recommendations for additional study, areas for review, application to your work environment, items for further discussion with associates.

Personal Action Items:	
Action Item	**Target Date for Completion**

Chapter Thirteen

PROJECT GRAPHICS

There are several forms of project graphics commonly used by project managers to provide information to project stakeholders or communicate status about the project. Graphics include drawings, illustrations, photographs, and charts. Common project graphics include the Gantt Chart, network diagram, PERT, PDM, learning curve charts, quality control charts, and S-Curves to illustrate project financial data such as earned value. Project graphics are used to enhance the reporting of project status and are directly related to the PMBOK® Guide Knowledge Area Project Communications Management.

PM Knowledge Note

Project graphical displays make project information easy to review and understand and are the prime means for tracking project cost, schedule, and performance. Properly used, project graphics will assist the project team in managing the project more effectively by clearly identifying variances and areas where attention or corrective action is required.

Glossary of terms Key terms and definitions to review and remember

Control Chart A tool used to determine whether a manufacturing or business process is in a state of statistical control or not. Generally includes a center line or process average an upper control limit and lower control limit. Data points are plotted on the control chart and then analyzed to determine where process problems may exist.

Flow Chart A pictorial summary that uses symbols and words to illustrate the steps, sequence, and relationships of the various operations in the performance of a function or process.

Gantt Chart A bar chart that depicts activities of a project as bars or "blocks" over time. The beginning and end of the bar or block correspond to the beginning and end date of the activity. A scheduling tool used to display the status of project activities. A graphical representation of a project that shows each task as a horizontal bar whose length is proportional to its time of completion. Named after Henry Gantt, Gantt charts are most commonly used to exhibit project or program progress. They do not effectively illustrate dependencies although today's project software provides the capability to display dependencies on a Gantt chart. The greatest advantage of a Gantt or bar chart is that it is easy to read and understand and can be changed easily and quickly.

GERT Chart   Graphic Evaluation and Review Technique. This type of chart allows for feedback loops and conditional branches. It may include decision points or possible alternative paths depending on specific conditions.

Graphics   Visual elements that supplement typed information to make printed messages more clear or interesting. Graphics are basically pictures and drawings. They may be created by computer or drawn by a graphic artist. Graphics are frequently created by the project team to support project documentation and ensure that the information provided is clearly understood.

Histogram   In statistics, a graphical display of tabulated frequencies, shown as bars

Ishikawa Diagram   Also known as Fishbone and Cause and effect diagram. Assists in determining the root causes of a stated problem.

Network Diagram   A graphical diagram that uses boxes to represent project activities and lines to show the relationship between activities. Generally, the sequence is shown from left to right, and the logical relationships between activities are shown with arrows or lines.

Pareto Diagram   A chart used to graphically summarize and display the relative importance of the differences between groups of data. Generally shown as a bar chart that indicates from left to right the largest cause of failure or frequency of occurrence of a cause associated with a problem or quality issue to the lowest level or frequency of failure or occurrence. The Pareto diagram is used to prioritize items that require attention.

PERT Chart   Program Evaluation Review Technique. A diagram that depicts project tasks and their interrelationships. Developed by the U.S. Navy in the 1950s, it is designed to schedule, organize, and coordinate activities within a project. The PERT Chart uses probabilistic estimates to determine project duration based on optimistic, most likely, and pessimistic estimates.

Activities, Questions, and Exercises

Refer to Chapter Thirteen of *Project Management: A Systems Approach to Planning, Scheduling, and Controlling* (10th Edition) for supporting information and assistance in completing each exercise.

The following questions and exercises are most closely associated with the knowledge areas of the *PMBOK® Guide*: Project Time Management, Project Communications Management

Review each of the following questions or exercises and provide the answers in the space provided. Some questions and situations may have more than one appropriate answer or response.

1. ✍ **PM Quick Check:**

   A. The Gantt chart is intended to effectively and clearly show project activity dependencies.

   True _____

   False _____

   B. Network diagrams are used to depict project status and progress.

   True _____

   False _____

   C. A Gantt chart and a bar chart are generally used for the same purpose.

   True _____

   False _____

   D. What is the major difficulty encountered when developing a network diagram?

   _____

   _____

2. Explain the benefits of using each of the following types of charts:

Chart	Advantage
1. Gantt Chart	
2. Network Diagram	
3. Organization Chart	
4. PERT Chart	
5. Flow Chart	
6. GERT Chart	
7. Pareto Diagram	

3. ✍ **PM Quick Check:**

Create a network diagram using the following information:

Task	Dependency
Start	None
A	Start
B	A
C	A, B
D	C
E	D
F	D, B
G	E, F
H	G, E
End	

4. Match the type of chart with its main intended purpose.

A. Gantt Chart		**1.** Displays logic and dependencies
B. GERT Chart		**2.** Displays process steps and sequences
C. Network Diagram		**3.** Indicates project progress
D. Pareto Diagram		**4.** Displays feedback loops, conditional branches, and alternative paths
E. Flow Chart		**5.** Displays causes in order of frequency of occurrence

5. A WBS is a tool used to:

a) Show project dependencies
b) Display the project schedule
c) Organize the project deliverables into specific groups
d) Define resource responsibility

6. The following spread sheet is known as a _____.

WBS Task	Project Manager	Sponsor	Functional Manager #1	Functional Manager #2

7. A WBS and an organization chart are very similar in their construction and appearance.

True _____

False _____

8.

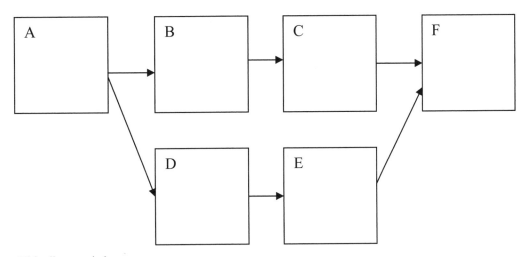

This diagram is known as a:

a) Activity-on-Arrow diagram
b) PDM or Activity-on-node
c) GERT Chart
d) Gantt Chart

PM Knowledge Note

Project graphics enhance communication between stakeholders and reduce the amount of written project documentation.

9.   True or False   The pareto diagram is a form of histogram.

True _____

False _____

10.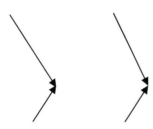

What is this diagram known as and what is its main purpose?

_____

_____

11.   A *histogram* is "a representation of a frequency distribution by means of bars whose widths represent class intervals and whose areas are proportional to the corresponding frequencies."
Create a histogram using the following data: Data set - {3, 11, 12, 19, 22, 23, 24, 25, 27, 29, 35, 36, 37, 45, 49}, 50, 52, 55, 58

Data Range	Frequency
0-10	
10-20	
20-30	
30-40	
40 -50	
50-60	

Making the grade! An exam tip for you to review.

> *Remember the different types of project graphics and their advantages and disadvantages. The Gantt Chart primarily shows progress; the network diagram primarily shows sequence and dependencies.*

PMP® & CAPM® Exam tip

Kerzner "Quick tips" for the Project Management Institute PMP® and CAPM® EXAM

The topic "Project Graphics" is most closely related to the following areas in the *PMBOK® Guide*: Project Time Management and Project Communications Management

Project graphics can enhance project communications by providing additional clarity to project documentation.

Identify the needs of the project stakeholders before preparing project charts and diagrams. The needs of project executives and sponsors are, in most cases, different from functional manager or customer needs. Project sponsors and executives typically expect to receive summary information about the condition of a project. Sponsors may be executives within an organization or they may be steering committees or groups of decision makers. Project managers are usually selected by the project sponsor or project executive.

Remember that each type of project graphic has a different intended purpose. Multiple graphics are often used to ensure that all information is provided in an understandable format.

> Additional tips and practice items for the PMP® exam are included in each chapter and in the section of the workbook entitled *PMP® Exam and PMBOK® Guide Review.*

Answers to Questions and Exercises

1. A—False, B—False, C—True, D—The network diagram does not address the question "What if something goes wrong?" There are no feedback loops or alternate paths in a network diagram.

2.

Chart	Advantage
1. Gantt Chart	Clearly shows progress of activities
2. Network Diagram	Displays dependencies
3. Organization Chart	Displays areas or responsibility, effectively displays chain of command, displays general communications flow
4. PERT Chart	Displays activity dependencies, weighted average estimates, project duration, and critical path
5. Flow Chart	Shows a complete process flow, allows for identification of gaps, very effective in communicating process information
6. GERT Chart	Allows conditional branches and feedback loops to address different possible outcomes
7. Pareto Diagram	Displays frequency of occurrence of a problem. Identifies and prioritizes items that warrant immediate attention

3.

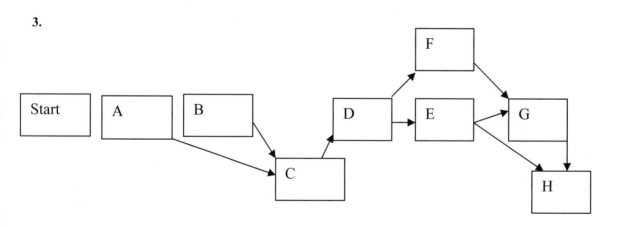

4.

A. Gantt Chart	3	1. Displays logic and dependencies
B. GERT Chart	4	2. Displays process steps and sequences
C. Network Diagram	1	3. Indicates project progress
D. Pareto Diagram	5	4. Displays feedback loops, conditional branches, and alternative paths
C. Flow Chart	2	5. Displays causes in order of frequency of occurrence

5. c

6. RAM

7. True

8. b

9. True

10. Ishikawa or Cause and Effect diagram – used to identify specific problems and determine potential root causes that can be further defined and resolved.

11.

Data Range	Frequency
0-10	1
10-20	3
20-30	6
30-40	4
40-50	3
50-60	4

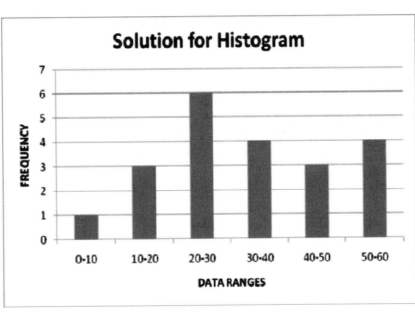

Your Personal Learning Library

Write down your thoughts, ideas, and observations about the material in the chapter that may assist you with your learning experience. Create action items and additional study plans to assist you in enhancing your skills or for preparing to take the PMP® or CAPM® exam.

Insights, key learning points, personal recommendations for additional study, areas for review, application to your work environment, items for further discussion with associates.

_____

_____

_____

Personal Action Items:	
Action Item	**Target Date for Completion**

Chapter Fourteen

PRICING AND ESTIMATING

There are many activities associated with the planning of a project. Estimating the likely costs of all resources and work effort required to complete the project is a significant part of the planning process. There are several common techniques available for use in estimating project costs, activity duration, and project completion date and the project manager may use a combination of these techniques throughout the life cycle of the project. These techniques vary in application but the objective of the estimating process is to determine, within a desired degree of accuracy, the total cost to produce a product or project deliverable. The difficulty of estimating costs varies widely depending on the type of project, the experience level of the project team, the number and type of resources to be used, and whether or not similar projects have been completed that can be used for comparison and analogy. Estimating is basically a guess or an approximation of project costs or time durations and includes a risk assessment and in most cases an added contingency to address uncertainty.

Pricing is generally the process of determining the appropriate amount to bill or charge a client, customer, or receiver of a product or service. Pricing is considered to be a business decision and is associated with profitability, relationship with clients, the general target market and what the market will bear, availability of the product, difficulty to produce, availability of materials, competition, and risk factors.

It is important to emphasize that one of the most crucial inputs in the pricing decision is the cost estimate of the proposed project baseline. Pricing actually begins before the proposal development process. It starts with customer requirements, well understood subtasks, and in many cases a top down estimate that is further refined during the planning process.

There are three major types of estimates that may be used during project planning as the project is executed: Analogous, Parametric, and Bottom up or grass roots estimating. The accuracy of each of these types must be considered and additional, more detailed and rigorous estimating may be required.

Generally, estimates are more accurate and reliable when they are developed by the functional organizations or specific resources that have been assigned responsibility for an activity. The experience and lessons learned from functional groups increase the reliability of an estimate and will in most cases improve the quality of the overall planning process.

PM Knowledge Note

Estimates are not based on luck. They are well-thought-out decisions based on the best available information, expert judgment, best practices, and some type of model that is based on previously collected and validated data.

Glossary of terms   Key terms and definitions to review and remember

Analogous Estimate   Preparing an estimate by comparing projects that may be similar to some degree. This type of estimate is considered to be a top down estimate and has a low degree of accuracy. Analogous estimates are in most cases refined to improve accuracy over time.

Bottom-up Estimating   A technique that uses well-defined project information, generally from a WBS, to develop an estimate with a high degree of accuracy.

Cost Baseline   The approved total estimated cost of the project. The cost baseline is the output of the cost-budgeting process. The cost baseline is a time-phased budget that is used as a basis for measuring and controlling project costs through each phase of the project life cycle.

Cost Budgeting   The aggregation of all project costs to produce the project cost baseline.

Cost Estimating Relationships   The output of cost models. Typical Cost Estimating Relationships include:

- Mathematical equations based on regression analysis
- Cost-quantity relationships such as learning curves
- Cost to cost relationships
- Cost–noncost relationships based on physical characteristics, technical parameters, or performance characteristics

Estimating   The process of calculating the likely cost of a project. It may be a formal or informal process. The act of determining the likely cost of construction works on behalf of either clients or contractors. Determining the probable cost of future work.

Estimating Groups   These are people or departments that estimate the cost of new work without soliciting input from the functional managers. The success of this group is based upon the quality of the estimating data base

Estimating Teams   On some projects, the initial group of people assigned to the project are functional subject matter experts with knowledge of functional estimating standards. Once the estimating team completes their tasks, they may be replaced by other functional employees who will be doing the actual work.

Feasibility Study   A process to determine if a business idea is capable of being achieved by an organization. A study of the applicability or ability to develop and deliver a proposed action or plan.

Forward Pricing Rates   Long term forecasts on the salary of the workers two or three years ahead. This is for estimating on long term projects. Forward pricing rates are also used on overhead costs and material costs.

Fully Loaded or Fully Burdened Rate   The total cost of a resource including salary and all overhead costs such as benefits. The total of all costs associated with maintaining a resource. Many organizations establish reference tables that provide the fully burdened rates of its employees for use in project cost estimating.

Management Reserve (as related to pricing)   A sum of money or a percent of the estimated cost that is added into the cost baseline for possible inaccuracies in the estimating process

Parametric Estimating   A process of estimating that utilizes mathematical models to determine an approximate outcome. An example would be the use of a standard cost per unit multiplied by the number of units required. Example: House construction may use a standard cost per square foot of living space multiplied by the desired amount of living space. This method is also referred to as a top down estimate.

Pricing   The process of applying prices to purchases and sales orders based on several factors: fixed amount, quantity, promotion or sales campaign, specific vendor quote. Pricing is the process of determining the correct amount to bill or charge for a product or service. Pricing includes the cost of developing and building or fabrication of a product, investment risks, and fair profit margin.

Regression Analysis   A method for determining the association between a dependent variable and one or more independent variables. The statistical technique of finding a straight line that approximates the information in a group of data points. A statistical method to estimate any trend that might exist among important factors.

Sunk Cost   This is a cost incurred in the past that will not be affected by any future decision. Sunk costs are not considered when making decisions about whether to continue with a project or terminate the project. A cost that has been incurred and cannot be recovered.

Top Down Estimating   This is an estimating process at the top one or two levels of the WBS. Estimating can be done quickly but the accuracy of the estimates may be risky unless the company has a good estimating data base.

Activities, Questions, and Exercises

Refer to Chapter 14 of *Project Management: A Systems Approach to Planning, Scheduling, and Controlling* (10th Edition) for supporting information. Review each of the following questions or exercises and provide the answers in the space provided.

The following questions and exercises are associated with the knowledge areas of the *PMBOK® Guide*: Project Cost Management and Project Time Management. There is also a relationship to Project Risk Management

PM Knowledge Note

The accuracy of estimating types is a key factor in the planning process. The type of project, especially where new technology is involved or where a new product is being developed, may require several adjustments to the prepared estimates as the project proceeds through progressive elaboration.

1. **Project Estimating Types**

Review each of the following estimating types and match the
descriptions and/or characteristics.

Estimating Method	Generic Type	WBS Relationship	Accuracy
Parametric			
Analogy (Analogous)			
Engineering or Grass Roots			

Generic Type—A. ROM (Rough Order of Magnitude)   B. Budget   C. Definitive

WBS relationship—A. Top down   B. Bottom-up

Accuracy—A. –25% to +75%   B. –10% to +25%   C. –5% to +10%

2. ✑ **PM Quick Check:**

A. Which of the following will provide the project team and project manager with the most useful
information to develop a well-prepared and definitive project cost estimate?

   a) The Project Charter
   b) The Preliminary Scope Statement
   c) The WBS
   d) The Statement of Work

B. The ROM or rough order of magnitude estimating type is associated with which of the following
estimating methods?

   a) Engineering or grass roots
   b) Parametric
   c) Analogous
   d) Definitive

C. Project activity estimates should be developed by:

   a) The project sponsor
   b) The project manager
   c) The functional manager
   d) The customer

> **PM Knowledge Note**
>
> *The WBS is considered the cornerstone of project planning and in addition to displaying the project tasks, activities and project complexity is used by the project team to increase the probability of producing reliable cost estimates, activity duration estimates, and better risk assessment!*

3. Explain how the WBS is used by the project manager and project team to develop a reliable bottom-up estimate.

_____

_____

4. What is the difference between a "sunk cost" and an "opportunity cost"?

_____

_____

5. **Direct Costs and Indirect Costs**

   Review the list of project costs and identify each item as a direct cost or an indirect cost.

Direct Cost	Indirect Cost

Project Costs:
A. Project manager travel expenses
B. Project team labor rates
C. Equipment costs
D. Taxes
E. Maintenance and janitorial services
F. Reward and recognition of the project team
G. Company benefits
H. Materials
I. Accounting services
J. Utilities

PM Knowledge Note

Variable costs and fixed costs should be considered when planning the project budget. Variable costs are costs that may change during the project life cycle including material costs, supplies, and salaries. Maintain an awareness of the use of "bargained for" resources that may be associated with contractual salary or wage increases during the project life cycle. Fixed costs generally do not change and may be associated with nonrecurring or recurring activities. Fixed costs may include rentals, one-time setup, or leases.

6. ✍ **PM Quick Check:**

PMP® & CAPM® Exam

A. Using the technique of analogous estimating is also associated with the technique known as:

a) Scope decomposition
b) Scope definition
c) Feasibility analysis
d) Expert judgment

B. Before developing the project cost estimates for resources the project manager should consider the following organizational expense items: building maintenance, fringe benefits, group insurance, office supplies, clerical support, postage, retirement plans, sick leave, vacation pay. These items are generally associated with:

a) Contractor direct costs
b) Overhead rates or costs
c) Variable costs
d) External costs

7. **Developing a Cost Estimate**

There is generally a sequence of steps taken to determine the estimated cost of a project deliverable. Review the following list and, using note paper or Post–It® Notes, place each step in the appropriate sequence to display the complete process.

A. Develop the WBS	
B. Provide a complete definition of the work requirements	
C. Estimate the costs (cost out) of the WBS	
D. Establish a logic network	
E. Establish reasonable costs for each WBS element	
F. Decide on the basic course of action	
G. Review WBS costs with functional managers	
H. Develop the RAM	
I. Negotiate with functional managers for qualified personnel	

J. Review base case costs with upper management or sponsor	
K. Document the results in the project plan	
L. Establish cost summary reports	
M. Develop the final detailed network diagram and CPM schedules	

8. ✎ **PM Quick Check:**

PMP® & CAPM® Exam

A. Sunk costs are used as a major factor in the decision to approve the continuation of a project into the next phase.

   True _____ False _____

PM Knowledge Note

In some cases the customer may wish to audit the cost proposal before it is approved. The project manager should prepare supporting data about cost estimates for reference during audits.

9. **Estimating Pitfalls**

For each of the following estimating pitfalls and problems provide a possible preventive measure or solution.

Pitfall or Problem	Recommended Solution
A. Misinterpretation of the statement of work	
B. Omissions or improperly defined scope	
C. Poorly defined or overly optimistic schedule	
D. Inaccurate WBS	
E. Applying improper skill levels to project activities	
F. Failure to account for risks	

G. Failure to understand or account for cost escalation and inflation	
H. Failure to use the correct estimating technique	
I. Failure to use forward pricing rates for overhead, general administrative, and indirect costs	

10. Review the following common risks that may be encountered when planning a project. These items may significantly impact the project cost or activity duration estimates and ultimately affect the final project results. For each risk identified provide a possible preventive action or response.

Risk	Response or Preventive Action
Poorly defined requirements	
Lack of qualified resources	
Lack of management support	
Inexperienced project manager	
Unreliable estimates from functional managers	
New technology or work that has not been performed previously	
Unreliable subcontractors	

11. Life-Cycle Costing

Which of the following costs are generally associated with the project life cycle, and which costs are associated with the total life cycle of the product?

A. R&D Costs   Project Life Cycle _____ Product Life Cycle _____

B. Production Costs   Project Life Cycle _____ Product Life Cycle _____

C. Construction Costs   Project Life Cycle _____ Product Life Cycle _____

D. Operation & Maintenance Project Life Cycle _____ Product Life Cycle _____

E. Product Retirement   Project Life Cycle _____ Product Life Cycle _____

PM Knowledge Note

Life cycle cost analysis is the systematic analytical process of evaluating various alternative courses of action early in the project with the objective of choosing the best way to employ scarce resources and lower the cost of the total life cycle of the product or the project.

☑ **Study Notes:**

Successful application of the life-cycle costing process will:
- Provide downstream resource impact visibility
- Provide more effective life-cycle cost management
- Influence/improve/enhance R&D decision making
- Support downstream strategic budgeting

Estimating methods may be classified as follows:

Informal
- Expert judgment or experience
- Analogy
- SWAG (basically a guess with no actual reliable basis)
- ROM (Rough Order of Magnitude)

Formal
- Engineering/Detailed from standards
- Parametric
- Organizational process standards and assets

12. Explain the advantages and disadvantages of each estimating type listed in the table.

Estimating Type	Advantages	Disadvantages
Engineering or Grass Roots (Bottom-up estimating)		
Parametric		
Analogous (Analogy)		

☑ **Study Notes:**

Economic Project Selection Methods

- **Payback Period**—The time required to recover or pay back an initial investment. Payback does not take into consideration the time value of money. Example: A project requires an initial investment of $10,000. The project team estimates recovery of the initial investment in two years. The payback period is two years. If this is acceptable to the organization decision makers, the project may be approved.

- **Discounted Cash Flow (DCF)**—Determining the present value of potential future earnings of an investment. The yield of the investment is calculated in terms of the present value. Example: If an investment yields $1,000 one year from now and the cost of money is 10%, the value in today's dollars would be $909. This is determined by the formula for present value: $PV = FV/(1 + r)^n$ where PV = Present Value, FV = Future Value, r = the rate or cost of money, and n = the number of years. The future value of an investment is determined by the formula $FV = PV (1 + r)^n$.

- **Breakeven Analysis**—The point at which the cash outflow and the cash inflow meet. The cash outflows are projected along with the cash inflows. The date where the two curves intersect is the breakeven point. If the breakeven point is acceptable to the organization, the project will be approved.

- **Net Present Value (NPV)**—Net Present Value is determined by calculating the present values for the cash flows for each year of the project. The sum of the present values is determined and the Net Present Value is calculated by subtracting the initial investment from the sum of the present values. Generally a Net Present Value equal to or greater than zero would indicate an acceptable project.

- **Internal Rate of Return (IRR)**—A sophisticated capital-budgeting process involving a series of trial and error or iterative calculations to determine at what interest rate or cost of capital rate would result in a Net Present Value of zero. This requires several calculations at increasing cost of capital rates to determine the actual Internal Rate of Return.

- **Depreciation**—The reduction in the balance sheet of a company asset to reflect its loss of value through age and wear. The decline in the value of a capital asset. Depreciation represents a cost of ownership and the consumption of an asset's useful life. Generally deductible for tax purposes.

13. ✐ **PM Quick Check:**

A. What is the future value of $5,000 invested at 10% interest for 3 years?

_____

B. What is the Present Value of a project that will yield $100,000 two years from now if the cost of capital is 15%?

_____

C. What is the Net Present Value of the following project?
The discount rate is 5%. The initial investment is $10,000 and the project has a 5-year duration.

Year	Cash Inflows	Present Value
1	$1,000	
2	$2,000	
3	$2,000	
4	$5,000	
5	$2,000	
	Sum of Present Values	Sum =
	Initial Investment (-)	
	Net Present Value	NPV =

14. A Problem to Consider: Estimating Person Day

Companies usually estimate work based upon man-months (person months). If the work must be estimated in person-weeks, the person-month is then converted to person-weeks. The problem is in the determination of how many person-hours per month are actually available for actual direct labor work.

Your company has received a request for proposal (RFP) from one of your customers and management has decided to submit a bid. Only one department in your company will be required to perform the work and the department manager estimates that 3,000 hours of <u>direct</u> labor will be required.

Your first step is to calculate the number of hours available in a typical person-month. The Human Resources Department provides you with the following <u>yearly</u> history for the average employee in the company:

- Vacation (3 weeks)
- Sick days (4 days)
- Paid holidays (10 days)
- Jury duty (1 day)

A. How many direct labor hours are available per month per person?

```
 2,080  hours available per year
 –120   for vacation
  –32   for sick days
  –80   for paid holidays
   –8   for jury duty
─────
 1,840  hours per year or 153 per month
```

B. If only one employee can be assigned to the project, what will be the duration of the effort, in months?

3,000 hrs/(153 hrs per month) = 19.6 months

(This assumes only one person)

C. If the customer wants the job completed within one year, how many employees should be assigned?

3,000 hrs/(1,840 hrs/person) = 1.6 people

Kerzner "Quick tips" for the Project Management Institute PMP® and CAPM® EXAM

The subjects in this chapter are most closely associated with the knowledge areas of the PMBOK® Guide: Project Cost Management and Project Time Management.

Remember the three types of estimating methods—Analogous, Parametric, and Grass Roots or Bottom-up estimating.

Analogy, or Analogous, and Parametric estimating are considered top down estimating techniques and are not generally very accurate.

Pricing is a business decision and is determined after project costs have been estimated and validated.

The cost baseline is the aggregate of all cost estimates and is provided to the sponsor for review and approval.

The cost baseline is allocated across each phase of the project to create a time phased budget that allows the project costs to be monitored as the project progresses through each phase.

Sunk costs are those costs that have been incurred and are not considered when making decisions about the future of the project.

Opportunity cost is generally considered the loss of revenue that is experienced when one project is selected over another or when funds are held for one project that could have been used to generate revenue in another area or on another project.

Remember to consider overhead and indirect costs when developing a project budget.

Life-cycle costing includes all costs from R&D to salvage. Project costing is generally associated with the costs to deliver the project and do not include operations, maintenance, and salvage.

> Additional tips and practice items for the PMP® exam are included in each chapter and in the section of the workbook entitled **PMP Exam and PMBOK Guide® Review**.

Answers to Questions and Exercises

1.

Estimating Method	Generic Type	WBS Relationship	Accuracy
Parametric	A	A	A
Analogy (Analogous)	B	A	B
Engineering or Grass Roots	C	B	C

2. A—C, B—B, C—C

3. The WBS provides the details of the project activities and allows the team to determine the resource requirements and other cost factors to produce a more accurate estimate.

4. A sunk cost is an expenditure that has been incurred while working on a project and cannot be recovered. It is not considered when making decisions about the future disposition of the project. Opportunity cost is associated with the loss of revenue that could have been attained from one project by holding funds for another project.

5.

Direct Cost	Indirect Cost
A	D
B	E
C	G
F	I
H	J

6. A—d, B—b

7.

A. Develop the WBS	3
B. Provide a complete definition of the work requirements	1
C. Estimate the costs (cost out) of the WBS	4
D. Establish a logic network	2
E. Establish reasonable costs for each WBS element	7
F. Decide on the basic course of action	6
G. Review WBS costs with functional managers	5
H. Develop the RAM	10
I. Negotiate with functional managers for qualified personnel	9
J. Review base case costs with upper management or sponsor	8
K. Document the results in the project plan	13
L. Establish cost summary reports	12
M. Develop the final detailed network diagram and CPM schedules	11

8. False

9.

Pitfall or Problem	Recommended Solution
A. Misinterpretation of the statement of work	Ensure the SOW is free of imprecise language. Have the SOW reviewed by a third party.
B. Omissions or improperly defined scope	Verify the scope statement and project scope. Use a template or check list to define the scope. Obtain support from functional groups.
C. Poorly defined or overly optimistic schedule	Ensure that functional groups are involved in the planning, consider risks associated with scheduling, validate assumptions.
D. Inaccurate WBS	Use functional manager knowledge, use templates from previous projects, review WBS with scope statement and statement of work.
E. Applying improper skill levels to project activities	Identify the appropriate skill levels required, provide training as needed, verify competency levels.
F. Failure to account for risks	Prepare a risk management plan and conduct risk analysis.
G. Failure to understand or account for cost escalation and inflation	Review lessons learned, obtain support from accounting groups, ensure that estimates include forecasts of economic conditions.
H. Failure to use the correct estimating technique	Review standard practices, review expectations with project sponsors and executives.
I. Failure to use forward pricing rates for overhead, general administrative, and indirect costs	Consider the time value of money, plan for inflation and changing market prices for material, identify overhead (rent, telecom costs, support services, benefits), consider contractural raises bargained for by employees, obtain templates to assist in estimating, obtain a third party review of estimates.

10.

Risk	Response or Preventive Action
Poorly defined requirements	Develop a plan to identify requirements early in the project. Verify statement of work and scope statement. Validate requirements before developing project plans.
Lack of qualified resources	Define resource requirements. Review WBS to identify resource needs. Negotiate with functional managers for qualified resources. Communicate resource requirements prior to project kickoff.

Risk	Response or Preventive Action
Lack of management support	Connect the project to strategic goals. Review the project objectives and scope statement with management. Verify that the project was selected using the appropriate selection process.
Inexperienced project manager	Request a more experienced project manager. Establish a mentor relationship with a more experienced project manager. Define required skills before selection of a project manager.
Unreliable estimates from functional managers	Obtain records and historical data from previous projects. Request supporting data for all estimates.
New technology or work that has not been performed previously	Prepare rolling wave estimates. Reassess estimates at the completion of each phase.
Unreliable subcontractors	Establish contractor selection criteria. Establish expectations and performance measurement criteria. Negotiate penalty clauses and other controlling processes in the contract.

11. A—Product Life Cycle, B—Project Life Cycle and Product Life Cycle, C—Project Life Cycle and Product Life Cycle, D—Product Life Cycle, E—Product Life Cycle

12.

Estimating Type	Advantages	Disadvantages
Engineering or Grass Roots (Bottom-up estimating)	Detailed information Very accurate Provides best estimating base for future project change estimates	Requires significant amount of time to complete May be costly
Parametric	Generally a simple process Statistical data base can provide expected values and prediction intervals	Requires the establishment of parametric cost relationships to be established (need a basis for the estimates) Considered top down and low in accuracy
Analogous (Analogy)	Low cost Quick Simple to generate Accurate when very similar projects are compared	Generally low in accuracy except when projects are very similar Limited to stable technology

13.

Year	Cash Inflows	Present Value
1	$1,000	952
2	$2,000	1,814
3	$2,000	1,728
4	$5,000	4,113
5	$2,000	1,567
	Sum of Present Values	Sum = 10,174
	Initial Investment (-)	10,000
	Net Present Value	NPV = 174

Your Personal Learning Library

Write down your thoughts, ideas, and observations about the material in the chapter that may assist you with your learning experience. Create action items and additional study plans to assist you in enhancing your skills or for preparing to take the PMP® or CAPM® exam.

Insights, key learning points, personal recommendations for additional study, areas for review, application to your work environment, items for further discussion with associates.

_____

_____

_____

_____

_____

_____

_____

_____

_____

_____

Personal Action Items:	
Action Item	**Target Date for Completion**

Chapter Fifteen

COST CONTROL

Project cost control includes the monitoring of project activities, the tracking and recording of project data, and the corrective actions taken to return project performance to acceptable levels when variances have gone beyond established parameters. Cost control processes should be introduced to the project team at the start of the project or at the kickoff meeting and should be performed by all project personnel who incur costs. Generally, cost control includes cost estimating, cost accounting, managing project cash flow, and other activities and tactics that may affect the financial aspects of the project. The project budget prepared by the project manager and project team includes the estimated cost of all project deliverables and the activities required to produce them plus any overhead costs, indirect costs, and contingencies that have been included to address potential project risk events. The project manager is held accountable for the successful management and control of project costs and must be prepared to act quickly when project costs begin to approach established thresholds. Continuous monitoring, analysis of variances, and identification of alternatives that may be utilized to correct variances are one of many project manager responsibilities.

The project manager's main objective regarding cost control is to maximize the probability of completing the project within the approved budget by keeping actual costs at or below planned levels, to minimize the use of available contingency reserves, to maximize company profits, or to minimize overall expenses (depending on the type of project).

It is important for the project manager to understand that cost management is directly related to schedule management and scope management. These three elements are associated with the Triple Constraint and awareness of the inter- relationship between these elements is an essential factor to achieving overall project success.

Study Notes: Some key activities of cost control:

- Awareness of factors that may cause changes to the cost baseline
- Obtaining approval of changes that will affect the cost baseline
- Managing changes through a change control process to prevent additional cost-related issues from occurring or to minimize the impact on other projects or company operations
- Monitoring cost performance on a regular basis through project reviews
- Recording and documenting all project changes that impact the cost baseline
- Preventing incorrect or inappropriate changes
- Informing project stakeholders of approved changes
- Taking corrective action to resolve identified cost overruns

PMP® &
CAPM®
Exam

Cost control also includes the use of effective risk management techniques that address the opportunities as well as the potential negative events that may be experienced during project execution.

PM Knowledge Note

Any cost control system is only as good as the original plan against which performance will be measured. Effective planning and the use of reliable sources of estimates are key factors in managing project costs.

Glossary of terms   Key terms and definitions to review and remember

Actual Cost (AC)   Amount expended (in terms of direct labor or direct cost) of work that has been completed within a given time period. Also known as ACWP, or Actual Cost of Work Performed.

Actual Cost of Work Performed (ACWP)   Amount expended (in terms of direct labor or direct cost) on work that has been completed within a given time period. Also known as AC, or Actual Cost.

Baseline   The original approved plan for a project. There are three major baselines associated with a project – The Scope baseline, the Schedule baseline, and the cost baseline or project budget..

Benefit-cost ratio (BCR)   An indicator, used in the formal discipline of cost-benefit analysis, that attempts to summarize the overall value for money of a project or proposal. A BCR is the ratio of the benefits of a project or proposal, expressed in monetary terms, relative to its costs, also expressed in monetary terms. All benefits and costs should be expressed in discounted present values

Budget at Completion (BAC)   The sum of the total budgets for a project. Sum of the budgets for each phase of a project.

Budgeted Cost of Work Performed (BCWP)   Amount budgeted or planned to be expended for work that has been completed. Also known as Earned Value.

Budgeted Cost of Work Scheduled (BCWS)   The budgeted amount of work that has been scheduled to be completed at the time of measurement. Also known as Planned Value.

Contingency   The planned allotment of time and cost for unforeseeable elements that may be associated with project planning. A planned reaction or response established to address an event that may or may not happen.

Corrective Action   Changes made to bring expected future performance of the project in line with the plan. Actions taken to return project performance to the planned level or performance or to an acceptable level.

Cost Control   The process of controlling the cost of a project within a predetermined sum throughout its various stages or phases.

Cost Performance Index   The cost efficiency ratio of earned value to actual costs. Expressed as the formula CPI = CV/BCWP or CV/EV CPI. It is often used to predict the magnitude of a possible cost overrun using the formula BAC/CPI = projected cost at completion.

Cost Variance   Any difference between the budgeted cost of an activity and the actual cost of that activity. In earned value BCWP – ACWP or EV – AC.

Earned Value (EV)   A measure of the physical work accomplished considering the original approved cost estimate or authorized budget for this work. The sum of the approved cost estimates (may include overhead allocation) for activities (or portions of activities) completed during a given period (usually measured to a specific point in time). Also known as Budgeted Cost of Work Performed, or BCWP.

Earned Value Management   A method for integrating scope, schedule, and resources, and for measuring project performance. It compares the amount of work that was planned with the work that was actually completed or "earned" with the units of cost expended (actually spent) to determine if cost and schedule performance are in line with the project plan. Also expressed as BCWP or EV.

Effort   The number of labor units required to complete an activity or other project element. Usually expressed as staff hours, staff days, or staff weeks.

Estimate   An assessment of the likely quantitative result. Usually applied to project costs and durations and should always include some indication of accuracy such as plus or minus a percent.

Estimate at Completion (EAC)   The expected total cost of an activity, a group of activities, or the project when the defined scope of work has been completed.

Estimate to Complete (ETC)   The expected remaining costs to complete an activity, a group of activities, or the project.

Planned Value (PV)   The physical work scheduled plus the authorized budget to accomplish the scheduled work. Also known as BCWS, or Budgeted Cost of Work Scheduled.

Schedule Performance Index (SPI)   The schedule efficiency ratio or earned value accomplished against the planned value. The SPI describes what portion of the planned schedule was actually accomplished. SPI = BCWP/BCWS or EV/PV.

Schedule Variance   Any difference between the scheduled completion of an activity and the actual completion of the activity. In earned value, BCWP – BCWS or EV – PV.

Scope Creep   The uncontrolled and unauthorized changes to the project scope. Usually results in greater cost, lower quality, and unfavorable performance variances.

Variance   Any deviation from the plan.

Work Breakdown Structure (WBS)   A deliverable-oriented grouping of project elements that organizes and defines the total scope of the project. Each descending level represents an increasingly detailed definition of the project work.

Work Package   A deliverable at the lowest level of the work breakdown structure.

Study Notes: EVM Formulas—A brief review and study tool

Special attention: PMI® refers to Earned Value (BCWP) as EV, Planned Value (BCWS) as PV, and Actual Cost (ACWP) as AC.

Formulas to remember:

CV or Cost Variance =	BCWP – ACWP	or EV – AC
SV or Schedule Variance =	BCWP – BCWS	or EV – PV
CPI or Cost Performance Index =	BCWP/ACWP	or EV/AC

SPI or Schedule Performance Index =	BCWP/BCWS	or EV/PV
CV % or Cost Variance Percent =	CV/BCWP	or CV/EV
SV % or Schedule Variance Percent =	SV/BCWS	or SV/PV

EAC or Estimate at Completion = ACWP + ETC or AC + ETC where ETC is the Estimate to Complete . This formula is used when the original estimates appear to be questionable or are no longer relevant due to changes in the project. (ETC refers to work that has not been performed and is therefore an estimate of cost.)

EAC = AC (cumulative) + BAC – EV (cumulative where BAC is the Budget at Completion, this formula is used when the variances experienced are atypical and similar variances are not expected to occur as work proceeds.

EAC = Ac (cumulative) + (BAC –EV Cumulative) divided by CPI (Cost Performance Index). This formula is used when variance experiences are typical and will continue to be observed as the project continues.

VAC or Variance at Completion =	BAC – EAC
Project Cost at Completion =	BAC/CPI

Cost – Benefit Ratio   The total saving or realized benefits in dollars (converted to present value) divided by the initial cost. Example: A program that cost $54,000 to develop and deliver resulted in a $430,000 saving the first year. 430,000 / 54,000 = 7.96. For every dollar spent on this investment there was a return of 7.96 dollars. This would be considered to be a very beneficial investment.

Attack of the Acronyms!

Review these acronyms and learn to recognize their meanings. These are the "language" of earned value. Learn to speak it fluently.

AC	Actual Cost
ACWP	Actual Cost of Work Performed
BAC	Budget at Completion
BCWP	Budgeted Cost of Work Performed
BCWS	Budgeted Cost of Work Scheduled
CPI	Cost Performance Index
CV	Cost Variance
CV %	Cost Variance Percent
EAC	Estimate at Completion
ETC	Estimate to Complete
EV	Earned Value
EVM	Earned Value Management
PMBOK®	Project Management Body of Knowledge
PV	Planned Value
SPI	Schedule Performance Index

SV	Schedule Variance
SV %	Schedule Variance Percent
VAC	Variance at Completion
WBS	Work Breakdown Structure

> **PM Knowledge Note**
>
> *The WBS is a key input in the cost management process. The WBS provides a basis for estimating the resources required, the duration of each activity, and for managing changes to the project plan. Project plans can be expected to change through the project life cycle but a well-planned project will generally have minimal scope changes. Ensure that the entire project team is involved in the development of the WBS. This will assist in developing a more complete project plan and improve how the team performs together. The WBS also allows the team to determine responsibility for each major task when it is associated with the RAM—Responsibility Assignment Matrix.*

Activities, Questions, and Exercises

Refer to Chapter 15 of *Project Management: A Systems Approach to Planning, Scheduling, and Controlling* (10th Edition) for supporting information. Review each of the following questions or exercises and provide the answers in the space provided.

The following questions and exercises are associated with the knowledge areas of the *PMBOK® Guide*: Project Cost Management and Integrated Change Control.

> **PM Knowledge Note**
>
> *The estimating process is the key to cost management. Make sure that your estimates have been obtained from reliable sources. Generally, estimates should be provided by the functional managers or subject matter experts that will perform the work. Remember that estimates are actually guesses or approximations and many factors can impact the accuracy of an estimate. Risk should be considered when developing estimates.*

1. Match Makers

A. In the left column are questions that can be answered by one and only one of the terms in the right column. Match the right column to the terms in the left column.

a) What is the revised estimate for the final cost at the completion of the project? _____

b) How much money was expended thus far? _____

c) How much physical work was actually accomplished thus far? _____

d) How much money was originally budgeted? _____

e) From where we are today, how much money is needed to complete the work? _____

A. PV
B. EV
C. AC
D. BAC
E. SV
F. CV
G. EAC
H. ETC

f) By how much have we deviated from the
schedule baseline, favorably or unfavorably?   _____

g) By how much have we deviated from the
cost baseline, favorably or unfavorably?   _____

h) What is the planned cost or value of the work
accomplished thus far?   _____

B. There are basically two types of reserves that project managers should become familiar with: management reserves – generally associated with unkown unkowns (issues that can not be anticipated or planned for) and contingency reserves generally associated with known unknowns (issues and risks about which some date for planning exists). For each of the two reserves in the left column, match ALL of the appropriate possibilities from the right column that are associated with the type of reserve.

a) Management reserve: _____

b) Contingency reserve: _____

    A. Labor rate escalations
    B. Estimating errors
    C. Overhead escalations
    D. Material price escalations
    E. Scope changes
    F. Workmanship errors
    G. Flood, natural disaster
    H. Previously documented schedule delays

C. Executives want the answers to two questions, at a minimum, during briefings.

a) Where are we today?   _____

b) Where will we end up? _____

Match the questions with each of the following terms. (For each of the terms below, which question will be satisfied using the term?)

A. PV
B. EV
C. AC
D. EAC
E. ETC
F. VAC
G. SPI
H. CPI

D. Shown below are three types of performance reports. Which terms from the right column appear in each report?

Progress report: _____

Status report: _____

Forecast report: _____

    A. PV
    B. EV
    C. AC
    D. EAC
    E. ETC
    F. SPI
    G. CPI
    H. SV
    I.   CV

2. The 50-50 Rule

Using the 50-50 Rule and the figure below, determine the value for PV, EV, and BAC.

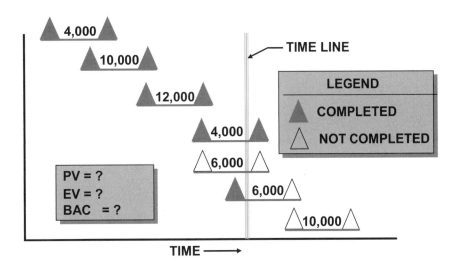

3. EVMS Problem: A customer (homeowner) hires a contractor to tile five identical rooms in his home. The customer purchased the tiles and the contractor will be reimbursed for labor only. Because the tiles are difficult to work with, the contractor assumes two days per room at eight hours per day and at $100 per hour. The planned cost is $8,000 but overtime will also be paid, if necessary, at the same rate of $100 per hour.

The first room was completed in three days because of difficulty in workmanship and getting an understanding of how to use these special tiles. This included two hours of overtime. The second room was completed in two days. Using EVMS, what information would be presented to the homeowner at the end of the first week?

a) PV = 5 x 800 = $4,000
b) EV = 4 x 800 = $3,200
c) AC = 42 x 100 = $4,200
d) BAC = $8,000
e) SV = EV – PV = –$800
f) CV = EV – AC = –$1,000
g) EAC = (AC/EV) x BAC = $10,500
h) VAC = BAC – EAC = –$2,500
i) ETC = EAC – BAC = $6,300
j) Percent complete = EV/BAC = 40%
k) SPI = EV/PV = 0.8
l) CPI = EV/AC = 0.76

Now it is your turn. Having tiled the floors, you now hire a contractor to wallpaper all of the walls in each of the five rooms. Once again, you purchase the wallpaper and will reimburse the contractor just for labor. The contractor works at $100 per hour and estimates eight hours of work per room, for a total of five days.

At the end of the first week, working eight hours per day, the contractor completed only three rooms. The EVMS status is:

a) PV = _____

b) EV = _____

c) AC = _____

d) BAC = _____

e) SV = _____

f) CV = _____

g) EAC = _____

h) VAC = _____

i) ETC = _____

j) Percent complete = _____

k) SPI = _____

l) CPI = _____

4. **Fill in the blanks:** By mistake, someone left spaces unfilled in the status report below. Please complete the report by filling in the missing data.

Activity	PV	EV	AC	SV	CV
a.	100	100	150	?	?
b.	?	200	?	90	0
c.	350	?	?	−100	?
d.	400	300	?	?	−50
e.	175	?	150	25	?
Totals	?	?	1,000	?	?

5. The Code of Accounts

The WBS below shows the code of accounts and costs for a given project. Every place a question mark exists, data is missing. Complete the chart.

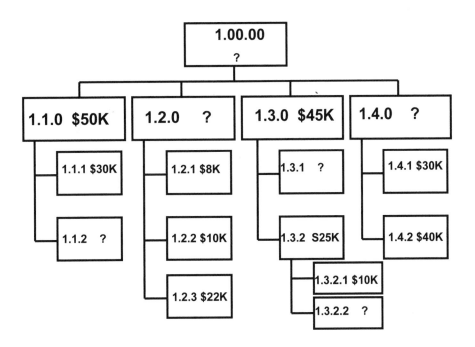

6. Understanding S Curves

Below are three S Curves. For each of the curves, determine whether the project is under budget, over budget, ahead of schedule, behind schedule, or cannot be determined.

A.

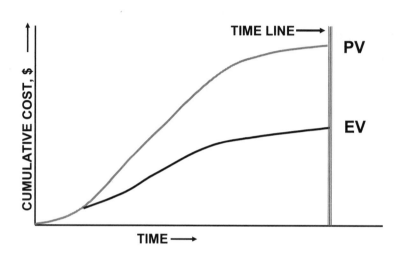

B.

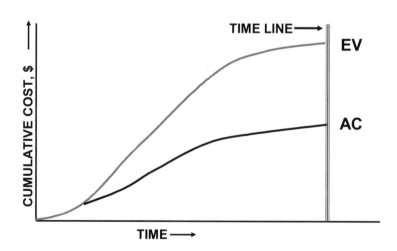

C.

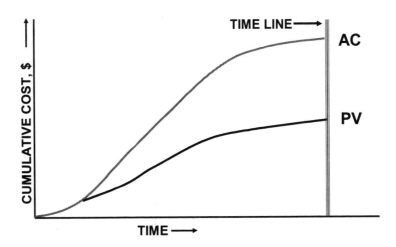

7. Understanding the Cost Breakdown

In the figure below, the project manager released $10 million to the functional areas to do the job. This was the released budget but the project manager also had, as part of the contract, $6 million for an undistributed budget that had not been planned out yet. The contract also has a $2 million profit included as well as a management reserve of $1 million. Using the figure, answer the following questions:

WBS Level 1 Cost Breakdown

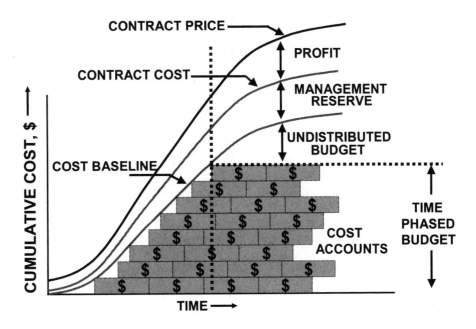

a) The dollar value of the cost baseline is: _____

b) The contracted cost is: _____

c) The contract price is: _____

8. **True-False:** Beside each statement, indicate whether the statement is true or false.

   1. If the cost variance is +$100, the activity is over budget.   _____

   2. CPI and SPI are used for trend analysis.   _____

   3. If SV = –$200, the project is over budget.   _____

   4. Management reserve is an example of an undistributed budget.   _____

   5. The most difficult part of EVMS is to determine EV or % complete.   _____

   6. EAC can be greater or less than BAC.   _____

   7. The 50-50 rule is used to determine AC.   _____

   8. Excluding management reserve and undistributed budgets, the summation of all of the PV for each work package should equal BAC.   _____

   9. Variances are usually expressed in percent as well as hours or dollars.   _____

   10. Scope changes should be paid for out of the management reserve.   _____

   11. Elements normally plotted in S curves are PV, EV, and AC.   _____

   12. As you progress through the life cycle phases of a project, the accuracy of the estimates usually improves.   _____

9. **You Know the Drill!**

   An Earned Value Drill to help you remember those formulas!

   Calculate the CV, SV, CPI, and SPI.

PMP® & CAPM® Exam

Case #	BCWS or PV	ACWP or AC	BCWP or EV	CV	SV	CPI	SPI
1	800	800	800				
2	800	600	400				
3	800	400	600				
4	800	600	600				
5	800	800	600				
6	800	800	1,000				
7	800	1,000	1,000				
8	800	600	800				
9	800	1,000	800				

(*continued on next page*)

Case #	BCWS or PV	ACWP or AC	BCWP or EV	CV	SV	CPI	SPI
10	800	1,000	600				
11	800	600	1,000				
12	800	1,200	1,000				
13	800	1,200	1,200				

A. Review the Earned Value analysis for case # 3. What is your assessment of this project?

_____

_____

B. Review the Earned Value analysis for case study # 12. What is your assessment of this project? What are some of the causes for this situation?

_____

_____

Kerzner "Quick tips" for the Project Management Institute PMP® and CAPM® EXAM

The subjects in this chapter are most closely associated with the areas of the *PMBOK® Guide*: Project Cost Management, Project Time Management, Project Scope Management.

Remember the basic earned value formulas and practice them on your actual projects. Use the technique to assess your project. Review the technique with your project teams and explain how the process applies to your project.

Remember that PMI® uses the Acronyms PV, AC, and EV. Many organizations continue to use the traditional acronyms BCWS, ACWP, and BCWP. Learn to use them interchangeably.

Earned Value Analysis is closely connected to communications management. The information calculated through earned value analysis is used to provide project stakeholders with useful information about a project's performance in the form of S Curves, charts, and status reports.

Cost control is a subset of Integrated Change Control.

Cost control is a subsystem of the Management Cost and Control System (MCCS).

Cost management is associated with cost estimating, cost accounting, direct costs, indirect costs, overhead, and many internal and external environmental factors.

Cost estimating should generally be performed by the functional groups or experts who will actually do the work.

Many projects require adjustments to the estimates as the project is progressively elaborated.

Practice developing S Curve charts to analyze project performance.

Project performance should be monitored on a regular basis.

Corrective action should be taken after alternatives have been identified. Select the best, most cost efficient trade-off.

Remember the relationships of the Triple Constraint.

Additional tips and practice items for the PMP® exam are included in each chapter and in the section of the workbook entitled **PMP® Exam and PMBOK® Guide Review.**

Answers to Questions and Exercises

1. Matching answers:

   A. a. EAC
      b. AC
      c. EV
      d. BAC
      e. ETC
      f. SV
      g. CV
      h. PV

   B. a. A, B, C, D, F, G
      b. E, H

   C. a. A, B, C
      b. D, E, F, G, H

   D. Progress report: A, B, C
      Status report: H, I
      Forecast report: D, E, F, G

2. The 50–50 Rule:
   PV = $34,000
   EV = $33,000
   BAC = $52,000

3. EVMS Problem

   a. $4,000
   b. $2,400
   c. $4,000
   d. $4,000
   e. −$1,600
   f. −$1,600
   g. $6,667
   h. −$2,667

i. $2,667
j. 60%
k. 0.60
l. 0.60

4. Fill in the blanks:

Activity	PV	EV	AC	SV	CV
a.	100	100	150	0	−50
b.	110	200	200	90	0
c.	350	250	150	−100	100
e.	400	300	350	−100	−50
f.	175	200	150	25	50
Totals	1,135	1,050	1,000	−85	50

5. The Code of Accounts:

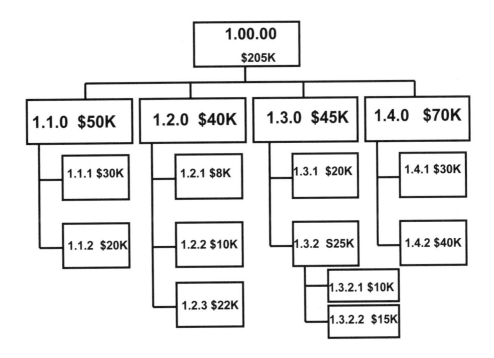

6. Understanding S Curves:

A. Behind schedule
B. Under budget
C. Cannot be determined

7. Understanding the Cost Breakdown:

A. Cost baseline is $16 million
B. Contracted cost is $17 million
C. Contract price is $19 million

8. True-False:

   1. False
   2. True
   3. False
   4. False
   5. True
   6. True
   7. False
   8. True
   9. True
   10. False
   11. True
   12. True

9. You Know the Drill! An Earned Value Drill to help you remember those formulas!

PMP® & CAPM® Exam

Case #	BCWS or PV	ACWP or AC	BCWP or EV	CV	SV	CPI	SPI
1	800	800	800	0	0	1	1
2	800	600	400	−200	−400	.66	.50
3	800	400	600	200	−200	1.5	.75
4	800	600	600	0	−200	1	.75
5	800	800	600	−200	−200	.75	.75
6	800	800	1,000	200	200	1.25	1.25
7	800	1,000	1,000	200	0	1	1.25
8	800	600	800	200	0	1.33	1
9	800	1,000	800	200	0	.80	1
10	800	1,000	600	−400	−200	.60	.75
11	800	600	1,000	400	200	1.66	1.25
12	800	1,200	1,000	−200	200	.83	1.25
13	800	1,200	1,200	0	400	1	1.50

A. Review the Earned Value analysis for case # 3. What is your assessment of this project?

It's good news and bad news. The work is being performed efficiently (under budget), but the project is behind schedule.

B. Review the Earned Value analysis for case study # 12.

The work is being done efficiently (ahead of schedule), but there is a cost overrun. Additional resources may have been used or overtime may have been used. There are several possible causes for this situation.

Your Personal Learning Library

Write down your thoughts, ideas, and observations about the material in the chapter that may assist you with your learning experience. Create action items and additional study plans to assist you in enhancing your skills or for preparing to take the PMP® or CAPM® exam.

Insights, key learning points, personal recommendations for additional study, areas for review, application to your work environment, items for further discussion with associates.

_____

_____

_____

_____

_____

_____

_____

_____

_____

_____

_____

Personal Action Items:	
Action Item	**Target Date for Completion**

Chapter Sixteen

TRADE-OFF ANALYSIS

Ideally, a project manager would like to complete a project on time, within budget, according to scope and quality requirements and achieve customer satisfaction. In a perfect project environment, this would be relatively easy to achieve if a well-thought-out plan was in place and the team was ready to handle any possible change or risk event that might possibly occur. In actual practice projects seldom progress exactly as planned. Customers request changes, the environment doesn't cooperate, people make mistakes, estimates are too optimistic, items are omitted or overlooked, conflicts occur, and team members are changed more frequently than we would like (churn or turnover of project teams can create rework, impede progress, and create conflicts). This "true" project environment drives the project manager and team to continually assess the project and, working within the identified and generally inflexible constraints, trade-offs (the giving up of one item to satisfy another need) must be considered. Project managers must be willing to analyze situations and decide on actions in a give-and-take project environment.

The time, cost, performance triangle, known as the Triple Constraint, is the "magic combination" that provides the basis for project trade-off decisions and keeps the project manager continuously pursuing a balance of these elements to achieve project success.

PM Knowledge Note

Although time, cost, and scope (or performance specifications) provide the basic success criteria, it is important to note that there are many other factors that must be considered in the quest to achieve customer satisfaction. Don't forget:
* *Providing what the customer needs*
* *Ensuring that the product is used*
* *The project should not interfere with normal business operations*
* *The project team sees personal value in completing the project*
* *Making sure the customer will want your organization to manage future projects*
* *The project should be perceived as supporting organizational strategic objectives*

Glossary of terms   Key terms and definitions to review and remember

Competing Trade-offs   More of one output requires less of another. Physical limitations limit the trade-off possibilities.

Complementary Trade-offs   Multiple outputs that can be achieved independently.

Constraint   A limitation or boundary.

Trade-off   A management decision to reduce one item in favor of another. A situation that requires a decision to give up one thing to obtain something else. An exchange that occurs as a compromise. A decision that implies that the decision maker fully comprehends both the positive (upside) and the negative (downside) of a particular choice.

Trade-off Analysis   A method to improve the quality of decisions by making the decision process more explicit, rational, and efficient. Generally a political process that can be structured in a systematic way to improve reliability and validity of the decision.

Triple Constraint   The three major baselines of a project that are interrelated. The triple constraint implies that any change to one of these elements will generally have an impact on one or both of the other elements. Usually depicted as a triangle.

Activities, Questions, and Exercises

Refer to Chapter 16 of *Project Management: A Systems Approach to Planning, Scheduling, and Controlling* (10th Edition) for supporting information. Review each of the following questions or exercises and provide the answers in the space provided.

The following questions and exercises are associated with the knowledge areas of the *PMBOK® Guide*: all knowledge areas, particularly Scope Management, Time Management, Cost Management, Quality Management, and Risk Management.

PM Knowledge Note	*There are two major elements of the trade-off analysis process:* • *Identifying and displaying how alternatives perform or affect a situation or defined criteria* • *Assisting stakeholders in articulating and applying their personal values to a problem in a rational way to achieve mutual agreement and a decision*

☑   **Study notes:** There are some generally accepted approaches in the trade-off analysis process:

Trade-off analysis is a systemic approach to managing the competing demands of time, cost and performance (scope or specifications). Information from cost reviews, schedule reports, and other forms of project reviews is compared to the original plan. This information can be used to support the following generally accepted activities associated with making trade-off decisions.

1. Engage the project team in a process oriented approach that will lead to consensus

2. Utilizing mathematical models and optimization calculations to determine the best available plan of action

3. Including key stakeholders in the process to gain support

4. Recognizing potential conflicts and determining their possible effects on the project

5. Reviewing and prioritizing project objectives

6. Analyzing the project environment and status of the project

7. Identifying alternative courses of action

8. Analyzing and selecting the best alternative

9. Obtaining approval to revise the project plan

Trade-off decisions may be difficult to reach. There are several factors that impact the decision process:

• Projects may be very complex with numerous system relationships
• The inherent uncertainty of many project environments
• The different opinions and perspectives of the project team and stakeholders
• Objectives may be unclear or conflicting
• Different priorities among functional groups

People make trade-off decisions every day. Consider the process of purchasing a new car. What are the trade-offs that would be considered?

Trade-off examples:

Price	Sound system
Safety	Security system
Maintenance costs	Domestic or foreign
Warranties	Fuel efficiency
Size	Resale value
Reliability	Durability
	Aesthetics

In the business environment trade-offs may be associated with: planning versus operations, projects versus daily activities and long-term programs, use of critical resources, funding one project over another, quality versus productivity, speed over accuracy. Other examples include Cost / benefit comparison, performance, response time, quality, schedule delays, scope reduction, choice of materials, choice of tools, choice of suppliers and resources.

Trade-off decisions require a method of measurement to effectively assess quantity- and quality-related issues and the impact to a project of one decision over another. Having specific metrics in place will also help to determine the actual impact of a decision after it has been implemented. Trade-off decisions may include several stakeholders including the project manager, sponsor, and the customer. Stakeholder bias and level of influence may significantly affect a trade-off decision.

When conducting trade-off analysis the project manager and other decision makers will assess the incremental cost of the trade-off, determine how it will reduce the problem or improve a situation, whether or not an alternative is worth doing, and rank alternatives from best to worst.

1. ✍ **PM Quick Check:**

   Explain the significance of the Triple Constraint and why it is depicted as a triangle.

   _____

   _____

2. **Making the Trade**

   You are the project manager for a project that is showing a schedule slippage, but the budget is under-running for the work that has been performed. What are the possible actions you may take to correct the schedule slippage? There are several possible answers.

   _____

   _____

   _____

3. The customer for your project insists on adding requirements that will increase the project scope. You are 50% complete with the project when the customer introduces a significant change in the configuration of a major deliverable. What actions may be taken to address this situation and what are the possible trade-offs that could be negotiated?

   _____

   _____

PM Knowledge Note

Decisions about trade-offs are generally based on the constraints of the project. It is not always possible to sacrifice cost, time, or performance without affecting the others.

4. **Constraints Alive!**

   What typical constraints may be encountered by a project team?

   _____

   _____

   _____

5. ✍ **PM Quick Check:**

Properly written objectives will help reduce the need for trade-off decisions. A well-written objective is characterized by the word _____.

There is generally a three-step process or methodology in trade-off analysis:

1. *Recognize and understand the conflict or issue. Make sure it actually requires attention.*
2. *Review the project objectives as seen by the participants. An understanding of objectives and priorities will determine the degree of rigidity in the decision process.*
3. *Analyze the project environment and status including financial risks, impact on other projects, impact on project results, customer perception.*

6. The project you are working on is over budget and behind schedule due to an unforeseen risk event. There is a contract clause that protects you from penalties but the customer is pressuring you to deliver on time. From the list provided, select the best options to consider.

Fast Track   _____

Reduce scope   _____

Crash the schedule   _____

Negotiate for more time   _____

Negotiate to phase in the deliverables by priority   _____

Terminate the contract   _____

Negotiate for more funding   _____

Revise the WBS and omit activities   _____

Shift critical resources to problem area   _____

7. Sometimes, the order of the trade-offs on the Triple Constraint is based heavily on the type of contract. For each of the contract types below, indicate which types the project manager would most likely allow cost overruns to occur before considering the other two items associated with the Triple Constraint.

   a) Cost plus percentage of cost _____

   b) Firm-Fixed Price_____

   c) Cost_____

   d) Cost sharing_____

   e) Cost plus incentive fee_____

   f) Cost plus award fee_____

   g) Cost plus fixed fee_____

8. Consider the following questions that may be asked by an executive during a project review. How would you respond?

   How can we get the project done faster?

   How can we complete the project at a lower cost?

How can you reduce the amount of overtime you are using?

How can we improve quality?

9. Your project schedule indicates that the work to be completed by a critical resource will not be completed in time to meet a predetermined release date for the resource. What are your options and potential trade-offs?

Possible Action	Trade -off
Example: Increase overtime for the critical resource	Additional cost, possible reduction in quality, burn-out of the resource,

Kerzner "Quick tips" for the Project Management Institute PMP® and CAPM® EXAM

The subjects in this chapter are most closely associated with the areas of the *PMBOK® Guide*: Project Integration Management, Project Scope Management, Project Time Management, Project Cost Management.

Trade-off analysis requires the project manager to communicate the issues as well as the potential impacts of a decision to the project stakeholders.

Effective planning will reduce the need to conduct trade-off analysis activities.

Ensure that an effective change control process is in place. Trade-offs will generally affect several areas.

The systems approach to managing projects emphasizes the importance of understanding that a change in one area, even if it is a very small change, can affect the entire project.

Make sure that objectives are clearly written and communicated to the stakeholders. Remember—objectives should be "SMART." (Specific Measurable, Attainable, Realistic Time Based)

Before making decisions about trade-offs, ask questions to determine why the trade-off decision is required and obtain information about the causes of the problem or the need for the trade-off. Consider the risks and potential consequences associated with trade-off decisions.

Document corrective action and the rationale behind trade-offs and record in a lessons-learned file or historical database.

The Triple Constraint provides a basis for understanding the effects of trade-off decisions related to schedule, cost, and scope. Changing the value of one element of the triple constraint will require some type of trade-off or adjustment to at least one other element

Trade –off analysis involves asking the question "How much must I give up to get more of what I want most?

Additional tips and practice items for the PMP® exam are included in each chapter and in the section of the workbook entitled *PMP® Exam and PMBOK® Guide Review.*

Answers to Questions and Exercises

1. The Triple Constraint generally refers to the relationship of three project baselines—Schedule, Cost, and Quality or performance specifications.

2. The first step would be to determine the severity of the slippage. It may be early in the project and the slippages may not be on the critical path. Overtime could be used or additional resources brought. The best approach is not to jump to conclusions or solutions. Obtain information about causes and then make a decision when you have all of the available data.

3. Start by ensuring that the change control process is followed. Identify and discuss the risks associated with the change request. Alternatives include: negotiating for more time and more resources, recommending the initiation of a new project after the completion of the current project.

4. Resource availability, imposed start and end dates, skill of the resources, contractual milestones, available funding.

5. SMART

6.

Fast Track	X
Reduce scope	X
Crash the schedule	X
Negotiate for more time	X
Negotiate to phase in the deliverables by priority	X
Terminate the contract	X
Negotiate for more funding	X
Revise the WBS and omit activities	X
Shift critical resources to problem area	X

7. Costs would most likely be sacrificed in a, c, d, e, f, g.

8. There are many possible responses. Also consider the tradeoffs associated with each decision.

How can we get the project done faster? – add more resources, reduce the project scope, bring in better more qualified resources.

How can we complete the project at a lower cost? – reduce scope, use less expensive resources, eliminate defects and poor quality

How can you reduce the amount of overtime you are using? – add more resources, add better trained and more efficient resources,

How can we improve quality? Provide training, conduct audits, use better materials, use trained resources

9.

Possible Action	Trade -off
Example: Increase overtime for the critical resource	Additional cost, possible reduction in quality, burn-out of the resource,
Locate additional qualified resources	Cost increase, other projects may be affected or delayed
Outsource the work	Cost, possible quality reduction, increased risk, reduced control of the work
Refuse to release the resource	Internal conflict with other project managers, escalation to management, potential problem obtaining resources in the future
Reduce scheduled performance tests	Poor quality, substandard performance
There are many additional possible actions and trade-offs. Consider your project and issues that may require trade-off decisions	

Your Personal Learning Library

Write down your thoughts, ideas, and observations about the material in the chapter that may assist you with your learning experience. Create action items and additional study plans to assist you in enhancing your skills or for preparing to take the PMP® or CAPM® exam.

Insights, key learning points, personal recommendations for additional study, areas for review, application to your work environment, items for further discussion with associates.

_____

_____

_____

_____

_____

_____

_____

_____

_____

Personal Action Items:	
Action Item	**Target Date for Completion**

Chapter Seventeen

RISK MANAGEMENT

Risk is a measure of uncertainty. In project management risk is generally considered to be a measure of the probability and consequence associated with a specific event or occurrence such as not achieving a defined project objective. Risk is also associated with potential opportunities that may enhance a project outcome as well as threats that may create serious project related problems during the planning and execution of a project. Risk is often viewed from a negative perspective and the focus of the project manager and team is on the harmful, undesired consequences, and expected losses that may occur. It is important to note that risk is also associated with the potential for beneficial results that may be experienced from project decisions. Risk management includes the processes associated with planning for risks, identification of risks, prioritization of risks through analysis, responding to the risks, and monitoring and controlling the project to prepare for new risks. The processes for risk management as described in the PMBOK® Guide—Fourth edition are:

- *Plan Risk Management*
- *Identify Risks*
- *Perform Qualitative Risk Analysis*
- *Perform Quantitative Risk Analysis*
- *Plan Risk Responses*
- *Monitor and Control Risk*

Risk management is an ongoing process that should be practiced in every phase of a project. The project team should become actively involved in risk management and adopt a "preventive" or proactive approach to managing risk. Project teams should include risk management plans in the early stages of the project and continue to emphasize the importance of regular risk reviews and assessments. An ideal point for a risk review is at the completion of a phase where lessons learned may be discussed and decisions about the next phase can be considered.

Project teams may develop well-written plans for guidance in execution but it is important to remember that plans are based on estimates, expert judgment, analogies, and lessons learned. Planning may reduce the probability of the occurrence of risk events and prepare the team to manage an event if it should occur but the project environment contains many uncertainties and therefore continuous emphasis on risk management is of considerable importance to the project manager and all stakeholders.

PM Knowledge Note

Risk has two primary components – Probability and Impact. Project Risk is associated with the uncertainty of possible events or conditions that, if they occur, may have a positive or negative impact on the project or the objectives of the project. Project Risks (potential occurrences) may have more than one cause and may result in the experience of more than one consequence. It is important to remember that there are different risks for each phase of a project and some risks may occur again. Continued emphasis about the importance of managing project risk and the need to have a plan in place to address the uncertainties associated with a project will increase the probability of successful project completion.

Glossary of terms   Key terms and definitions to review and remember

Brainstorming   A group creativity technique designed to generate a large number of ideas for the solution to a problem

Business Risk   A risk that may result in either profit or loss. Business risks are associated with competitor activities, inflation, recession, and customer response.

Contingency   The planned allotment of time and cost for unforeseeable events that may impact a project. An alternative or response to a realized threat.

Delphi Technique   A way to obtain the opinion of experts without necessarily bringing them together face to face.

Expected Value   The product of probability (P) × impact (I) or consequence.

Impact   The affect resulting from a realized threat or a defined vulnerability.

Insurable Risk   Also known as "pure" risk. A risk that will only result in loss—fire, theft, personal injury, direct property damage, legal liability

Maximin Criterion   Also known as Wald Critirion. The decision maker is concerned about loss and views risk from a pessimistic perspective.

Maximax Criterion   Also known as the Hurwicz Critirion. An optimistic approach to risk where the decision maker is optimistic about the results and attempts to maximize the benefits of a decision.

Monte Carlo Simulation   A method of generating values from a known distribution. Using random variables to produce failure times. A computerized technique that uses sampling from a random number sequence to simulate characteristics or events or outcomes with multiple possible values. Monte Carlo simulation is used to generate probable outcomes of a project based on available data that is processed many thousands of times to provide probable project results and information for project decision making.

Probability   A measure of how likely an event is to occur. A number expressing a ratio of favorable cases to the whole number of cases. Chance, odds, occurrence of uncertain events.

Reserve   Something kept back or saved for a special purpose. An amount of assets placed aside to respond to future needs or claims.

Risk   A measure of uncertainty. An exposure to chance of loss or damage.

Risk Averse   Avoids risks, concerned mainly with the possibility of loss or damage.

Risk Event   A discrete occurrence. The manifestation of risk into a consequence.

Risk Management   The act or practice of dealing with risk. Planning, identifying, analyzing, responding, and controlling project risks.

Risk Register   An organized list of risk events generally arranged by priority. Risk registers may include risk event details, probability and impact assessments and possible responses.

Risk Symptoms   Warning signs that a risk event may occur. Also known as Risk Triggers. Awareness and response to risk symptoms will reduce the probability of experiencing a risk event.

Utility   The ability of a stakeholder to accept and manage risk as the intensity and consequence of risk rises. The utility of a risk averter rises at a decreasing rate as risk increases. The utility of a risk seeker increases as risk increases

Workaround   A temporary response to an unplanned event.

Activities, Questions, and Exercises

Refer to Chapter 17 of *Project Management: A Systems Approach to Planning, Scheduling, and Controlling* (10th Edition) for supporting information. Review each of the following questions or exercises and provide the answers in the space provided.

The following questions and exercises are associated with the knowledge areas of the *PMBOK® Guide*: Project Risk Management.

PM Knowledge Note

Proper Risk Management is proactive rather than reactive. The project manager should instill risk management awareness and risk management practices within the project team at the project kickoff. Every team member should maintain awareness about risk and participate in risk management planning.

Risk Management Processes—A brief review

The following processes are generally associated with the knowledge area Project Risk Management. These processes are shown as sequential although in practice many of these process actions can be performed concurrently. It is important to remember that Risk Management begins with project planning and remains an ongoing process through each phase of the project life cycle. Refer to the *PMBOK® Guide*—Fourth edition for more information about the details associated with each process

Plan Risk Management

- The process of deciding how to approach and conduct risk management activities.

- **Input:** Enterprise environmental factors, organizational process assets, cost management plan, schedule management plan, communications management plan, project scope statement
- **Tools and Techniques:**
  - Planning meetings and analysis
- **Output:** The Risk Management Plan

Identify Risks

- The process of determining which risk events might occur and affect the project.
- **Input:** Enterprise environmental factors, organizational process assets, scope baseline, the risk management plan, activity cost estimates, activity duration estimates, stakeholder register, cost management plan, schedule management plan, project documents. Note that risk identification includes many of the project knowledge areas, several subsidiary plans, and an assessment odf stakeholders. Risk management is a group of integrated processes
- **Tools and Techniques:**
  - Documentation reviews
    Subsidiary plans
    Contracts
    Scope statement
    Statement of work
  - Information gathering techniques
    - Brainstorming
    - Delphi technique
    - Interviewing
    - Root cause identification
    - SWOT analysis
  - Checklist analysis
  - Assumptions analysis
  - Diagramming techniques
    - Cause and effect diagrams
    - System or process flow charts
    - Influence diagrams
  Expert Judgment
    Use of functional managers and subject matter experts
- **Output:** The risk register—This is a list of risks that includes risk categories, the identified risks, probability and impact of the risk, list of potential responses, root causes of risk, and the person assigned to manage the risk.

Perform Qualitative Risk Analysis

- The process of reviewing and prioritizing risk events and determining the probability and corresponding impact on project objectives. Generally uses lessons learned and expert judgment.
- **Input:** organizational process assets, project scope statement, the risk management plan, and the risk register.
- **Tools and Techniques:**
  - Risk probability and impact assessment
  - Probability and impact matrix
  - Risk data quality assessment—determining the reliability of the data source and the accuracy of the data obtained
  - Risk categorization—organizing risks by source
  - Risk urgency assessment—determining which risks have the greatest priority and must be addressed first

Expert Judgment
- **Output:** Updated risk register. The relative ranking or priority list of the identified risks, risks are grouped by category, a list of risks that require response in the short or near term, list of risks that require additional analysis, watch lists for low-priority risks, trends in qualitative analysis results that can be used to assist in forecasting potential impacts to the project.

Perform Quantitative Risk Analysis

- The process of reviewing and prioritizing risk events that have been identified and prioritized during qualitative analysis but require further consideration and more rigorous analysis.
- **Input:** organizational process assets, the risk management plan, the risk register, cost management plan, schedule management plan.
- **Tools and Techniques:**
  - Data gathering and representation techniques
    - Interviewing
    - Risk probability and impact assessment
    - Probability Distributions—Normal, Beta, Triangular
    - Expert judgment
  - Quantitative Risk Analysis and Modeling Techniques
    - Sensitivity analysis—determines which risks have the greatest potential impact on the project from an opportunity and threat perspective
    - Expected monetary value—the average outcome of a future scenario
    - Decision tree analysis—describes the possible implications of available choices
    - Modeling and simulation—the project is simulated many times to assist in determining possible outcomes. A common technique is Monte Carlo simulation.
- **Outputs:** Updates to the risk register, probabilistic analysis of the project, the identified probability of achieving project objectives, and a prioritized list of quantified risks

Risk Response Planning

- The process of developing options and approaches to the identified risks.
- **Input:** The risk management plan, and the risk register.
- **Tools and Techniques:**
  Strategies for negative risks or threats
    - Avoid—eliminating the risk
    - Transfer—shift responsibility to a more qualified or prepared resource or organization
    - Mitigate—reduce the probability of the occurrence and impact of the risk
    - Strategies for the positive impact of risk or opportunities
    - Exploit, share, or enhance
    - Acceptance—passive and active acceptance
  - Contingent response strategy—preparing for an event and responding when it occurs. (If this happens, we will respond to it.)
  - Expert Judgment
- **Outputs:** Risk register updates, risk related contract decisions project management plan updates, project document updates

Monitor and Control Risks

- The process of continually observing, tracking, analyzing the project for risk triggers, new risks, and reoccurrence of previously identified risks.

- **Input:** The project management plan, the risk register, work performance information (earned value analysis), performance reports.
- **Tools and Techniques:** Risk reassessment, risk audits (done periodically throughout the project), variance and trend analysis, technical performance measurement (to ensure that results meet specifications), reserve analysis (how the reserves established for the project are being utilized), status meetings to communicate information to stakeholders.
- **Output:** Risk register updates, change requests, recommended corrective actions, recommended preventive actions, updates to organizational process assets, project management plan updates, project document updates

1. ✎ **PM Quick Check:** Complete the following statement by placing the appropriate letter in the space provided.

The process of qualitative analysis is _____, while quantitative risk

analysis is_____.

A. a rapid and cost effective method of prioritizing identified risks
B. a method to determine the reliability of project risk data
C. used to develop risk-related decisions using mathematical analysis and simulations that present possible outcomes of risk events
D. used to create a watch list of low-priority risks

> **PM Knowledge Note**
>
> *A very effective tool for identifying project risks is the Work Breakdown Structure. The WBS displays the project activities and the complexity of the project and can be used by the project team to identify and then prioritize risks associated with each major deliverable or at the work package level. The work breakdown structure can be used to create a Risk Breakdown structure which will enable the team to categorize project risks.*

2. A Decision Tree Problem

An author decides to write a workbook about how to pass the PMP® exam on the first try. Historical data indicates that, if the market is strong, gross sales will be $1 million. If the market is weak, the gross sales will be only $50,000. There is a 70% chance of a strong market and a 30% chance of a weak market.

The author must decide whether to use a reputable publisher for the book or whether to self-publish the book himself. If he self-publishes the book, all of the sales receipts go to the author. But if he uses an external publisher, the publisher pays a 10% royalty on sales.

The decision tree for the problem is shown below. Usually, decision trees are prepared from right to left rather than left to right. In the diagram below, the boxes on the right represent the revenue that comes to the author, whether it be royalties or gross sales.

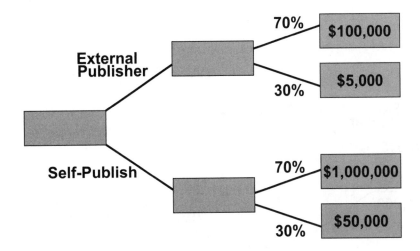

Now working from right to left, we can fill in the next row of boxes, which indicates the expected value of using the publisher as opposed to self-publishing:

Expected value (publisher) = 70% x $100,000 + 30% x $5,000 = $71,500

Expected value (self-publish) = 70% x $1M + 30% x $50,000 = $715,000

This is shown in the following figure:

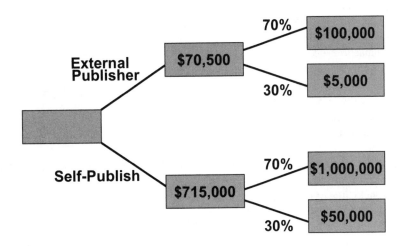

At this point, it certainly looks like you are better off self-publishing. But suppose your cost of self-publishing (i.e., materials, printing, marketing, advertising, etc.) is $600,000. This is an expense that must be subtracted from your revenue of $715,000. The result is shown in the figure below:

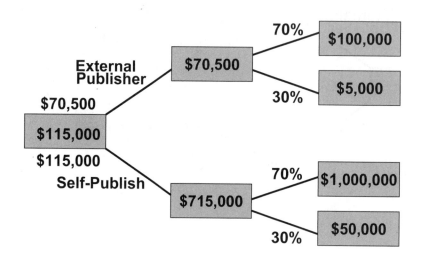

In the box on the left, you see the two options at this point: $70,500 income if you go with the publisher, and $115,000 if you self-publish. Therefore, $115,000 will go in the box on the left and, based upon probability theory, the best decision is to self-publish. While mathematically this might appear as the right decision, there is still another consideration. What happens if the author self-publishes and the market is weak? The author will receive $50,000 but may have expenses in the hundreds of thousands of dollars. The author can then lose a considerable sum of money. So, even though the solution says to self-publish, there is still a concern over the author's tolerance for risk.

2a. Decision Tree Analysis. You are considering the decision to purchase a machine for internal production or to subcontract the work to an external source. The following information has been provided by your financial managers:

PMP® & CAPM® Exam

Cost to purchase the machine—$35,000
Cost to subcontract the work—$5,000
Probability of a good market = 70%
Probability of a poor market = 30%
Reward if the prediction occurs:
In the purchase machine decision good market scenario—$80,000; in the poor market scenario—$30,000
In the Subcontract decision good market scenario—$50,000; in the poor market scenario—$15,000

1. What is the expected value of the decision to purchase the machine? _____

2. What is the expected value of the decision to subcontract the machine? _____

3. What is the most favorable decision based on the data provided by the financial managers?

3. Risk Tolerance

Explain the effect of a project that is experiencing progressively increasing risk on a person or an organization that is considered to be risk averse.

_____

_____

4. Risk Response Matching

Below are four responses to a risk situation made by a project manager. For each of the project manager's responses, from the list below select which response mechanism best fits the project manager's remarks.

- Assumption/ acceptance
- Avoidance
- Mitigation
- Transfer

1. "I know we have three different designs we can select from in order to satisfy the customer's requirements. However, I am very reluctant to consider the second option because, if we are wrong, there exists the chance of imminent damage to the project."   Response mode: _____

2. "This design has a very high possibility of creating a severe financial problem for us. I have asked the design team to develop an alternative that will minimize our risk exposure"

Response mode: _____

3. "I know we can manufacture the product ourselves, but there are a number of potential risks to consider. I would rather award a fixed-price contract to Alpha Company and have them do the manufacturing instead. They have a strong record of success in this area"
Response mode: _____

4. "I have set up a management reserve just in case there are escalations in the costs of the raw materials we must purchase." Response mode: _____

5. "There is a strong possibility that one of my team members may leave the project for a better paying job assignment so I have added an end of project bonus to retain the employee. Response mode

_____

6. "I have arranged for an insurance company to handle any direct property damage and any worker injuries during the project" Response mode _____

7. "We have assessed the risk and are aware that it may occur within the next phase of the project. We have developed a contingency to manage the risk if it occurs.

5. Contractual Risks

Every contract type can be viewed as a sharing of risks between the buyer and the seller. For each type of contract identified below, indicate who has the greater degree of risk: the buyer or the seller.

1. Cost plus percentage of cost: _____
2. Fixed price with economic price adjustments: _____

3. Fixed price incentive fee: _____

4. Cost plus incentive fee: _____

5. Time and materials not to exceed a certain amount: _____

6. Cost plus fixed fee (fee fixed in $, not %): _____

7. Cost plus award fee (fee is a % of a dollar pool): _____

8. Firm fixed price: _____

9. Cost sharing: _____

6. Risk Identification Matching

For the design and development of a new plane, Boeing identifies four categories of risks, which are shown in the left column below. In the right column are four possible mitigation strategies for the four categories. Select one and only one mitigation strategy from the right column to go with the left column.

1. Financial Risks: _____

2. Marketing Risks: _____

3. Technical Risks: _____

4. Manufacturing Risks: _____

a. Extensive testing of new parts before they go into a next generation plane. Testing could result in a next generation plane or a new plane.

b. Having firm orders for several planes before manufacturing begins.

c. Asking suppliers to share in the development costs for a new plane.

d. Offering a family of planes (i.e., 737, 747, 767, 777, 787).

7. True-False

1. One of the purposes of risk management is to create an understanding of the potential risks and their effects. True _____ False _____

2. Risk management is expensive and should be done on only large projects or those of strategic importance. True _____ False _____

3. Risk management is designed to provide an early warning system that will be useful throughout the project life cycle. True _____ False _____

4. All risk response mechanisms transfer all or part of the risk to a potential contractor. True _____ False _____

5. The best way to classify risks is according to their probability of occurrence. True _____ False _____

6. Risk management is designed to focus only on the potential negative situations and project threats that have been identified by the project team. True _____ False _____

7. A risk management plan should provide some guidance on how to respond or contain the risk event, if possible. True _____ False _____

8. A PI (probability-impact) Matrix can be used for either quantitative or qualitative assessment of risks. True _____ False _____

9. The Delphi Technique is a risk mitigation strategy whereby a group of subject matter experts work closely together in a two-day meeting to identify and solve risk problems. True _____ False _____

10. Expected value calculations are examples of quantitative risk assessment.
True _____ False _____

11. Planning is based upon history whereas risk management focuses on the future.
True _____ False _____

12. Risk and knowledge are inversely related. True _____ False _____

13. The major difference between each type of contract is the sharing of risk between the buyer and the seller. True _____ False _____

14. "Traffic light" or "Dashboard" identification of risks is a quantitative risk assessment method.
True _____ False _____

15. Triggers are early warning signs that a potential risk might be materializing.
True _____ False _____

PM Knowledge Note	*Project managers must be capable of managing in an uncertain environment. The project environment will change continuously and a proactive approach to risk management will help to minimize surprises and prepare the project team to respond effectively to risk situations. Remember, risk management is not a one-time event. It is applied through the entire life cycle of the project.*

8.   COOL TOOL!

Here's an effective tool for the project manager's toolbox: The project plan is actually an integrated plan that includes elements of all nine knowledge areas of the *PMBOK® Guide*. Create a checklist for managing risk by listing all nine knowledge areas and then, working with the project team, identify the risks that may be encountered in each knowledge area. This technique will help the project team develop a broader understanding of the project and further illustrate how a project actually works as a system.

Knowledge Area	Potential Risks	Response or Preventive Action
Integration		
Scope		
Time		
Cost		

Quality		
Human Resources		
Communications		
Risk		
Procurement		

9. Decisions, Decisions!

If you were asked to make a decision about how to respond to a project situation and you needed the opinions of several subject matter experts to review before you made a decision, which of the following techniques would be most effective if you wanted to keep the identities of each of the subject matter experts confidential to ensure that there was no possibility of influence or bias from other experts?

a) Nominal group technique
b) Brainstorming
c) Delphi technique
d) SWOT analysis

10. Going for the MAX!

A company with strong assets and a very aggressive approach to risk taking is most likely going to be associated with the _____.

a) Maximin or Wald criterion
b) Maximax or Hurwitz criterion

11.  **PM Quick Check:**

A. Inflation is an example of a business risk..

True_____ False_____

B. Bad weather is an example of insurable risk.

True _____ False _____

C. Utility is a measurement of risk tolerance

True _____ False _____

D. Probability x Impact = Expected value

True _____ False _____

12. What are the tree response modes for managing negative risks and project threats?

_____

13.   Consider the following situation. Your team has assessed the probability of experiencing software errors in the development process as medium and assigned a numerical value of the probability as .8. The team has also determined that the impact of these errors will also be high and have assigned a value of .9. What is the risk rating for the software error risk?

14.   What is the probability of rolling a one with one die (half a pair of dice?)

15.   What is the sum of all probabilities that can occur in a given set of circumstances?

16.   You are the project manager for the development of a new product. The following information has been provided to you by your analysts. There is an 80% chance of finishing on time and a 40% chance of finishing over budget. What is the probability that the project will finish on time and within budget?

17.   Consider the following project information provided by a functional manager: Optimistic time for completion of the project is 12 weeks. Pessimistic time is 40 weeks. Most likely time is 25 weeks.

    Using the weighted average formula, what is the probability that the project will be completed between 14.33 weeks and 33.01 weeks?

Kerzner "Quick tips" for the Project Management Institute PMP® and CAPM® EXAM

The subjects in this chapter are most closely associated with the area of the *PMBOK*® *Guide*: Project Risk Management.

Risk management should be practiced throughout the life cycle of the project.

Maintain an awareness of the risk tolerances of the organization, executive management, the customer, and the project team. A risk averse organization will attempt to minimize any exposure to potential risk events.

Identify sources of risk. Use brainstorming, Delphi technique or nominal group technique to identify potential risk situations.

Risks may be categorized to assist the team in identifying risks and also in preparing to respond to risks. An effective method for categorizing risks is to use the knowledge areas of the *PMBOK*® *Guide* as a basis.

Risk response is generally associated with four types of action: Avoidance, Transfer, Mitigation, and Acceptance.

There are two types of acceptance when responding to risk—passive or no action until the event occurs and active—planning a contingency to be ready if an event occurs.

As risks occur it is helpful to document the causes of the risk event and the rationale behind the corrective action that will be taken. The documentation can be placed in historical files and shared as lessons learned.

Remember: Risk will continue to appear throughout the project life cycle. Risk reviews and audits should be scheduled on a regular basis and risk management should be included in your project status meeting agendas.

Remember that the sum of the probabilities of each branch in the chance mode of a decision tree will always add up to 100%.

Additional tips and practice items for the PMP® exam are included in each chapter and in the section of the workbook entitled *PMP® Exam and PMBOK® Guide Review.*

Answers to Questions and Exercises

1. A, C

2a. 1. $65,000 2. $39,500 3. Subcontract the work

3. The utility or "satisfaction" associated with risk rises at a decreased rate as risk increases. Another way to explain this is that as risk increases the person or organization becomes more uncomfortable with the environment.

4. Risk Response Matching

1. Avoidance
2. Assumption/acceptance
3. Transfer
4. Mitigation
5. Mitigation
6. Transfer
7. Assumption/acceptance

5. Contractual Risks

1. Buyer
2. Seller
3. Seller
4. Buyer
5. Seller
6. Buyer
7. Buyer
8. Seller
9. Buyer

6. Risk Identification Matching

   1. c
   2. d
   3. a
   4. b

7. True-False

   1. F
   2. T
   3. F
   4. T
   5. F
   6. F
   7. T
   8. T
   9. F
   10. T
   11. T
   12. T
   13. T
   14. F
   15. T

8. This table is partially completed to provide examples of how the project team can effectively identify risks in each knowledge area and then develop possible responses to the risk. There are many possible answers and approaches to the risks identified in each knowledge area. Create your own template and use it as you manage your current or next project assignment.

Knowledge Area	Potential Risks	Response or Preventive Action
Integration	Failure to plan from an integrated or systems perspective	Utilize the entire project team developing the plan. Maintain awareness of the triple constraint
Scope	Omitted deliverables, incomplete scope statement, unclear objectives, scope creep	Clearly define objectives before planning begins, ensure that a complete scope statement has been prepared and has been verified, establish a change control process

9. c

10. b

11. A—True, B—False C. - True D. – True

12. Avoidance, Mitigate, Transfer

13. .8 x .9 = .72

14. 1/6 or 16.7 percent. There are 6 possible outcomes: 1,2,3,4,5,6. The probability of rolling a one is expressed as 1/6. (One of six possible outcomes)

15. 1.0. The sum of all probabilities will equal 1. Using the rolling of one die as an example there are 6 possible outcomes. : 1/6 + 1/6 +1/6 +1/6 + 1/6 + 1/6 = 1.0

16. You are the project manager for the development of a new product. The following information has been provided to you by your analysts. There is an 80% chance of finishing on time and a 40% chance of finishing over budget. What is the probability that the project will finish on time and within budget?

80% x 60% = 48%. There is a 60% chance the project will finish within budget. (40% chance of being over budget = 60 % of being within budget.

17. Consider the following project information provided by a functional manager: Optimistic time for completion of the project is 12 weeks. Pessimistic time is 40 weeks. Most likely time is 25 weeks.

Using the weighted average formula, what is the probability that the project will be completed between 14.33 weeks and 33.01 weeks?

Mean = 23.67 (optimistic + 4 times the most likely + pessimistic) / 6

Standard deviation = 4.67 (pessimistic – optimistic) / 6

Probability of completing the project between 14.33 weeks and 33.01 weeks = 95%

Mean of 23.67 minus 2 standard deviations = 14.33. The mean plus two standard deviations = 33.01. In normal distribution the probability of achieving a result with two standard deviations is 95%.

Your Personal Learning Library

Write down your thoughts, ideas, and observations about the material in the chapter that may assist you with your learning experience. Create action items and additional study plans to assist you in enhancing your skills or for preparing to take the PMP® or CAPM® exam.

Insights, key learning points, personal recommendations for additional study, areas for review, application to your work environment, items for further discussion with associates.

_____

_____

_____

_____

_____

_____

_____

_____

_____

_____

Personal Action Items:	
Action Item	**Target Date for Completion**

Chapter Eighteen

LEARNING CURVES

Learning curves were first introduced in 1936 by T.P. Wright. Wright described a theory for obtaining estimates based on repetitive production of airplane assemblies. The concept of learning curves has since been used for estimating all types of projects and programs from manufacturing to the space shuttle program. Learning curves are also known as progress functions. The theory of learning curves is based on the concept that repeating the same process or operation many times results in a reduction in the time and effort required to complete the process. Most people experience learning curves when they try a new product for the first time, when a new software application is used, or a new tool. The first time the process or operation is tried may take a considerable amount of time if it had never been performed before. The more the process, operation, or tool is used or performed the better the operator becomes at working with it. This reduces overall effort and decreases the actual time to complete the operation. The learning curve theory as written by T.P. Wright states that the direct labor hours necessary to complete a unit of production will decrease by a constant percentage each time production quantity is doubled. If the rate of improvement is 20% between doubled quantities, then the learning percent would be 80% (100 − 20 + 80). Learning curves are used to emphasize time but the principle can be related to cost also.
(Information adapted from the National Aeronautics and Space Administration—Cost Estimating Processes.)

Learning curves are most useful for estimating projects or operations that involve developing and estimating large quantities of deliverables or units. Determining the cost of producing the first unit may be relatively easy to calculate but the challenge is to determine the cost of the 100th, 1,000th, or 10,000th unit. The learning curve process is used to determine those costs and becomes a useful estimating tool.

Learning curve analysis is primarily used in situations that provide an opportunity for improvement or reduction in labor hours per unit.

PM Knowledge Note

Learning curves stipulate that manufacturing man-hours (specifically, direct labor) will decline each time a company doubles its output.
There are three basic conclusions about the learning curve theory:
- *The time required to perform a task decreases as the task is repeated*
- *The amount of improvement decreases as more units are produced*
- *The rate of improvement has sufficient consistency to allow its use as a prediction tool*

Glossary of terms   Key terms and definitions to review and remember

Cumulative Average Hours   The average hours expended per unit for all units produced through any given unit. When illustrated on a graph by a line drawn through each successive unit, the values for a cumulative average curve.

Cumulative Total Hours   The total hours expended for all units produced through any given unit. The data plotted on a graph with each point connected by a line form a cumulative total curve.

Slope of the Curve   A percentage figure that represents the steepness (constant rate of improvement) of the curve. Using the unit curve theory, this percentage represents the value (example, hours or cost) at a doubled production quantity in relation to the previous quantity. For example, with an experience curve having an 80% slope, the value of unit two is 80% of the value of unit one. The value of unit four is 80% of the value of unit two. The value of unit 1,000 is 80% of the value of unit 500.

Unit One   The first unit or product actually completed during a production run.

Unit Hours   The total direct labor hours expended to complete any specific unit. When a line is drawn on a graph through the values for each successive unit, the values form a unit curve.

A Review of PROJECT ESTIMATING USING LEARNING CURVES

Learning curves stipulate that the more often you perform a task, the less time (and money) is needed. This concept is very important to companies that have labor-intensive product lines with large volume production. Learning curves allow a company to take advantage of economies of scale and prevent new competitors from entering into the marketplace.

Learning curves have the following properties:

- The time required to perform a task decreases as the task is repeated
- Manufacturing man-hours will decrease by a fixed percentage each time production is doubled
- The rate of improvement has sufficient consistency to allow its use as a prediction tool

The following figure shows a typical learning when plotted in Cartesian coordinates:

Standard Learning Curve

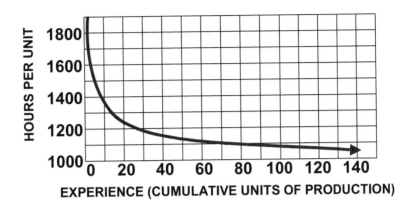

As the number of units produced increases, the time required for each unit decreases. Most companies use hours on the Y-axis rather than cost because cost can change resulting from salary, overhead and raw material price increases.

Most companies plot learning curves on log-log paper rather than in Cartesian coordinates. When this is done, the curve will appear as a straight line as shown in the following figure.

Standard Learning Curve

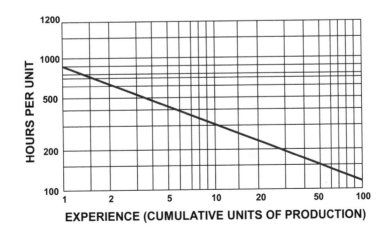

If the figure represents an 80% learning curve, then each time production doubles, the rate of learning will be at the same 80%. As an example, if the 100th unit produced required 300 hours, then the 200th unit produced would be 80% times 300 hours, or 240 hours. The 400th unit produced would then be 80% of the time needed for the 200th unit, or (80% x 240 hours), which equals 192 hours.

Learning curves are more appropriate for labor-intensive product lines rather than automated product lines because people can learn whereas capital equipment does not learn. An 80% learning curve is better than a 90% learning curve.

Learning curves can undergo improvements. In other words, there are activities that can help us go from an 85% learning curve to an 80% learning curve. Some of these activities include:

- Training and educating the workers
- Making improvements in the way the work is performed
- Installing new production processes
- Hiring more skilled labor
- Using higher quality raw materials that are easier to use
- Redesigning the product so that difficult manufacturing steps can be eliminated or minimized
- Offering workers incentives or disincentives to work better and smarter

There are also factors that can prevent improvements to a learning curve. These include:

- People refuse to learn better ways of doing their job
- People are not motivated
- People have no faith in the learning curve concept
- People will not work better or smarter without incentives
- People are expected to work in the way that violates health and safety protocols

The following figure shows what happens when the learning curve improves. This is often called "toe down" learning.

Toe Down Learning Curve

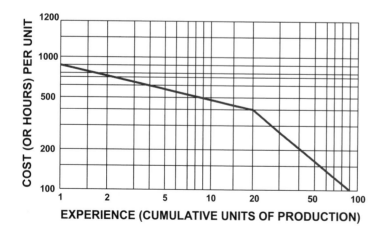

Learning curves work best if there is no break in the learning process. This means no long-term shutdown in the manufacturing process. For example, if the manufacturing process must be shut down for several months, perhaps because of large inventories and fewer customers, the learning curve will not start up at the same point because the workers may need to be re-educated on how to do the job. This is often referred to as a "toe up" learning curve as shown in the following figure.

Toe Up Learning Curve

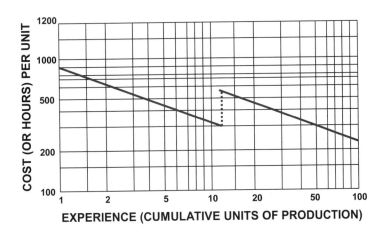

Learning cannot continue forever. Sometimes learning becomes "pegged" at a specific level as shown in the following figure. The pegged level becomes the standard for estimating the time and cost of producing a unit.

Pegged Learning Curve

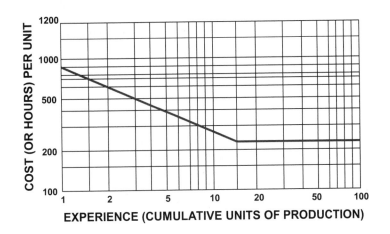

Learning Curve Problem: Shown below is the data for a 75% learning curve used in a manufacturing company.

Cumulative Units	Cost for This Unit ($)	Cumulative Cost ($)
1	616	616
2	462	1078
40	134	8467
100	92	14932
200	69	22751
300	58	29043
400	52	34513
500	47	39442
600	44	43980
700	41	48217
800	39	52212

A company manufactures products on a 75% learning curve. The company has already produced 100 units, which had a total cost of $14,932 according to the table.

A. What is the average cost per unit for the first 100 units?

$14,932 / 100 = $149.32 per unit

B. How much did it cost to produce the 100th unit?

From the table, the 100th unit was $92

C. According to learning curve theory, if production doubles to 200 units, the 200th unit should require only 75% of the 100th unit. Is that correct?

75% x $92 = $69, which matches the entry in the above table

D. If you have already produced 100 units and you have a customer that wants to purchase another 100 units, what is your average cost per unit for that customer for the next 100 units (i.e. units 101-200)?

($22,751 - $14,932) / 100 = $78.19 per unit

E. Sometimes, companies prefer to use hours rather than dollars. Assume in the table that the second and third columns represent hours rather than dollars. If the company has 1500 hours of manufacturing available per month to go from 101 to 200 units of production, how many months will be required?

(22,751 hours – 14,932 hours) = 7,819 hours

7,819 hours ÷ 1500 hours/month = 5.2 months

F. Whenever hours are used, we need to know the fully burdened employee labor rate to convert to dollars. Assume that the employees are fully burdened at $110 per hour and, using part E above, the company wishes to make a profit of 10% on the sale of units 101-200. Also, assume that 160 hours of management and supervision are need each month in addition to the manufacturing hours, and these hours are also billed at $110. (Disregard any costs for raw materials since this information is not provided.)

Manufacturing hours (from Part E) =   7,819 hours
Support hours = (5.2 months x 160 hours per month) =   832 hours
                                                                                    =========
                                                                                    8,651 hours

8,651 hours x $110 per hour = $951,610
10% profit =   95,161
                                                 ========
                                                 $1,046,771

G. Learning curves can also be used to price out products. As an example, using the table at the beginning of the problem, assume that the last two columns are costs rather than hours. Your company has produced 400 units thus far, and next year expects to sell 200 units according to marketing data. If these numbers are correct, and management wants to keep the price fixed for the entire year, using the learning curve for pricing what should be the fixed selling price of the product assuming a 10% profit margin is desired? This is an example of mid-point pricing.

The cost for the 400th unit is $52 per unit.

The cost for the 600th unit is $44 per unit.

Using midpoint pricing, the selling price = ($52 + $44) ÷ 2 = $48

Adding in a 10% profit, $48 x (1.10) = $52.80

One more problem.

You have already produced 200 units, and a new customer appears wanting to purchase 600 units. If your material cost is fixed at $60 per unit, a fully burdened hour is $110 and the above table represents labor hours only, how much should you charge the customer if you wish to make a 10% profit? (i.e. total cost) Also assume 1500 hours of manufacturing labor available per month and management support is 160 hours per month for the duration of the project.

Answer: Manufacturing labor hours for 600 units (unit 201 – 800) =
                    (52,212 – 22,751) = 29,461 hours

              Support hours = (29,461/1500) x 160 = 3,142 hours

              Labor cost = (29,461 + 3,142) x $110   = $3,586,330

              Total material cost = 600 units x $60   = $ 36,000
                                                                                    $3,622,330

                    10% profit   =   362,233
                                                                                    ========
                                                                                    $3,984,563

Now, let's add a small twist to this problem. Assume that learning is pegged when the 500th unit is produced. In other words, for all units produced after the 500th unit is manufactured, the hours needed for production of each unit will be at the hours needed for the 500th unit. How much time is required to produce units 201 – 800?

Units 201 – 500 require: 39,442 hours – 22,751 hours = 16,691 hours

From the table, the 500th unit (the pegged point) = 47 hours

Units 501 – 800 require: 300 units x 47 hours/unit = 14,100 hours.

Therefore, total hours = 16,691 + 14,100 = 30,791 hours

Activities, Questions, and Exercises

Refer to Chapter 18 of *Project Management: A Systems Approach to Planning, Scheduling, and Controlling* (10th Edition) for supporting information. Review each of the following questions or exercises and provide the answers in the space provided.

The following questions and exercises are associated with the knowledge areas of the PMBOK® Guide: Project Time Management and Project Cost Management.

1. **Learning Curve Problem:** Shown below is a 75% learning curve used in a manufacturing company.

Cumulative Units	Cost for This Unit ($)	Cumulative Cost ($)
1	616	616
2	462	1,078
40	134	8,467
100	92	14,932
200	69	22,751
300	58	29,043
400	52	34,513
500	47	39,442
600	44	43,980
700	41	48,217
800	39	52,212

A company manufactures products on a 75% learning curve. The company has already produced 100 units, which had a total cost of $14,932 according to the table. Questions and solutions have been provided for review purposes.

A. What is the average cost per unit for the first 100 units?
   $14,932/100 = $149.32 per unit
B. How much did it cost to produce the 100th unit?
   From the table, the 100th unit is $92
C. According to learning curve theory, if production doubles to 200 units, the 200th unit should require only 75% of the 100th unit. Is that correct?
   70% x 92 = $69, which matches the entry in the above table
D. If you have already produced 100 units and you have a customer that wants to purchase another 100 units, what is your average cost per unit for that customer?
   ($22,751 – $14,932)/100 = $78.19 per unit

Now it is your turn.

You have already produced 200 units, and a new customer appears wanting to purchase 600 units. If your material cost is fixed at $60 per unit and the above table represents labor hours only, and you wish to make a 10% profit, how much should you charge the customer? (i.e., total cost)

☑ **Study notes:** Calculating Learning Curves—T.P. Wright approach

The process calculates the average unit value of a production lot

$$Y = AX^b$$

Where:
Y = Cumulative average unit value of the Xth unit
A = The theoretical first unit value (T1)
X = Unit Number
b = Log (slope/Log 2)

For more information about learning curves, read chapter 18 of Kerzner, *Project Management—A Systems Approach to Planning, Scheduling, and Controlling* (9th Edition) (Wiley, 2006) or visit www.jsc.nasa.gov/bu2/learn.html.

Key point—The steeper the curve, the easier the learning.

Example of learning curve in action: It may cost $100 million to build the first copy of a new airplane, $80 million to build the second, $64 million to make the fourth, $51 million to make the eighth, and so on. The planes actually become less expensive to build as the workers and the company learn how to build them more efficiently. As the work is repeated, the workers become faster at completing tasks and they make fewer mistakes while reducing the amount of scrap and wasted material. (From Gary H. Anthes, *The Learning Curve: A Quick Study* in "Computer World," July 2, 2001.)

The very basic concept of the learning curve is that practice improves performance.

Learning Curve Limitations

- The learning curve does not continue forever. The percentage decline in hours/dollars diminishes over time
- The learning curve knowledge gained on one product may not be extendable to other products unless there exist shared experiences
- Cost data may not be readily available in order to construct a meaningful learning curve. Other problems can occur if overhead costs are included with the direct labor cost, or if the accounting codes cannot separate work packages sufficiently in order to identify those elements that truly demonstrate experience effects
- Quantity discounts can distort the costs and the perceived benefits of learning curves
- Inflation must be expressed in constant dollars. Otherwise, the gains realized from experience may be neutralized
- Learning curves are most useful on long-term horizons (years). On short-term horizons, benefits perceived may not be the result of the learning curves
- External influences, such as limitations on materials, patents, or even government regulations, can restrict the benefits of learning curves
- Constant annual production may have a limiting experience effect after a few years (example—no growth)

Kerzner "Quick tips" for the Project Management Institute PMP® and CAPM® EXAM

The subjects in this chapter are most closely associated with the areas of the _PMBOK®_ _Guide_: Project Time Management and Project Cost Management.

The basic principles of learning curves are:

1. The time required to perform a task decreases as the task is repeated.

2. The amount of improvement decreases as more units are produced.

3. The rate of improvement has sufficient consistency to allow its use as a prediction tool.

The learning curve theory is based on the principle that practice improves performance.

The consistency of improvement in the learning curve has been found to exist in a constant percentage reduction in time required over successively doubled quantities or units produced.

The constant percentage by which costs of doubled quantities decrease is called the rate of learning.

Additional tips and practice items for the PMP® exam are included in each chapter and in the section of the workbook entitled **_PMP® Exam and PMBOK® Guide Review._**

Answers to Questions and Exercises

1. Average labor cost for 600 units (unit 201–800) = ($52,212 – $22,751)/600 = $49.10 per unit

Total labor for these 600 units = 600 x $49.10 = $29,461

Total material cost = 600 units x $60   = $36,000
                                     ————
                                     $65,461

10 % profit   = $ 6,546
                                     ————
                                     $72,007

Your Personal Learning Library

Write down your thoughts, ideas, and observations about the material in the chapter that may assist you with your learning experience. Create action items and additional study plans to assist you in enhancing your skills or for preparing to take the PMP® or CAPM® exam.

Insights, key learning points, personal recommendations for additional study, areas for review, application to your work environment, items for further discussion with associates.

_____

_____

_____

_____

_____

_____

_____

_____

_____

_____

Personal Action Items:	
Action Item	**Target Date for Completion**

Chapter Nineteen

CONTRACT MANAGEMENT

Determining what items or resources must be obtained or acquired to successfully complete project deliverables and then negotiating the contracts that will guide the buyer and seller through the execution process can be very challenging regardless of the size or complexity of a project. In the contract and procurement environment there are two views, the buyer and the seller. These two views are often opposing and require effective negotiation to achieve mutually agreeable terms and conditions.

The Project Contract Management process includes the activities that will be used to determine what to purchase, how to purchase materials or resources, the preparation of documents that will be used in the process including RFPs (requests for proposal), selecting the appropriate suppliers, completing the contract and managing the contract, and closing or terminating the contract.

Contracts create legal obligations for the buyer and seller. These obligations include risks that must be considered before contractual terms are agreed upon. The risks for the buyer may be different than those of the seller depending on the type of contract that has been developed. Contract types are selected based on the needs of the buyer, the amount of detailed information that is available about the project, the deliverables that must be produced, the environmental factors that may affect the project, and the ability of the supplier to meet the needs of the buyer. A significant amount of negotiating may take place before a contract is agreed upon. The time and effort associated with contract negotiations and the impact of those negotiations on critical time frames must also be considered.

Major Contract Elements

1. **Intention to enter into a legal relationship.** Contracts are generally considered legal relationships and are presumed to include agreements about business, commerce, or other factors that are intended to be legally binding.

2. **Contracts include an offer and an acceptance.** The offer is a response to a need and the acceptance is based on the specific details of the offer.

3. **Consideration.** Consideration is something of value, usually money that will be exchanged for the service or product provided.

4. Capacity. The parties involved in the contract must have the actual capacity to fully understand and carry out the terms and conditions of the contract. A minor would be considered to lack the "capacity" to enter into a contract.

5. Genuine consent. Consent must be given without pressure or misrepresentation. The nature of the contract must be fully understood. If the behavior of the seller is such that a contract is agreed upon that has been misrepresented or forced due to the behavior of the seller, the contract may be subject to be voided.

6. Legality of objects. The contract must meet certain requirements for actual trade. There cannot be any terms that may actually be in violation of conditions established in the geographic area where the agreement has been negotiated. Restraints of trade must be considered before the contract terms and conditions are agreed upon.

PM Knowledge Note

In general, companies provide services or products based on the invitations for competitive bids issued by clients. One of the most important factors in preparing proposals and estimating the cost and profit of a project is the type of contract expected. There are contract types that may reduce the risk to the buyer and there are contract types that reduce the risk to the seller. Determining contract type will depend on the specific needs of each party involved in the negotiating process.

Procurement Strategies

There are two basic procurement strategies:

- Corporate procurement—the relationship of specific procurement actions to the corporate strategy. This is generally associated with a centralized procurement process to ensure the greatest benefit to an organization through control and economies of scale.

- Project procurement strategy—the relationship of specific procurement actions to the operating environment of the project. This strategy is focused at the project level and is focused on speed of acquisition and meeting the specific needs of the project.

Procurement planning generally involves the following decisions or objectives:

- Procure all goods and services from a single source
- Procure all goods and services from multiple sources
- Procure only a small portion of goods and services
- Procure none of the goods and services (all production will take place internally)
- Procure goods and services from a sole source—only one supplier is available and there is no choice but to use that supplier

Procurement management also includes the "make-or-buy" decision process, the rent-or-lease decision, the selection of suppliers, negotiation of the contract terms, selection of contract type, and the administration of the contract throughout the project life cycle. At project closeout the contract is reviewed to determine if all terms and conditions including all agreed upon deliverables have been produced and are accepted.

Glossary of terms Key terms and definitions to review and remember

Acceptance Unconditional agreement to an offer. The agreement creates the contract as solid and binding on both sides. Prior to signing, any party is free to withdraw.

Arbitration An independent third party used to settle disputes out of court. The arbitrator must be agreed upon by both sides. Contracts can have clauses nominating arbitrators in advance in case the need arises.

Bidder Conference A meeting during which information is shared by the buyer to all sellers who have an interest in presenting a bid or proposal.

Collective agreement Term for agreements made between employees and employers. Usually involves trade unions. These are often covered by more than one organization and can be seen as contracts but are not governed by contract law.

Condition A term of fundamental importance to the contract, the breach of which may cause the contract to be terminated.

Consideration In a contract each side benefits in some way, they give each other some consideration. This would usually be a price paid by one in exchange for goods supplied by the other. But anything of value to the other party can be used.

Contract Any legally enforceable promise or set of promises made by one party to another. An oral or written agreement between two or more parties which is enforceable by law.

Design Specification The specific details in terms of physical characteristics that describe, define, and specify what is to be done. The risk of performance is on the buyer.

Force Majeure Exemption from penalties due to non-fulfillment of contractual obligations resulting from conditions beyond the control of the parties involved. Examples—earthquake, floods, war.

Functional Specification A subset of performance specification. The specific end use of the product. What actions or functions it is intended to do when placed into use.

Indemnity Generally, a payment or compensation for damages done. Protection against future loss. An obligation of one party to reimburse another party for losses which have occurred or which may occur.

Liquidated Damages A sum, usually a fixed amount per day, to be paid as damages to an owner by a contractor due to failure to complete the specified work within the timeframe stipulated in the contract.

Make or Buy The decision to make a product or produce a service internally or purchase the required item from an external supplier.

Performance Specification The measurable capabilities the product must achieve in terms of operational characteristics. The risk of performance is on the contractor.

Privity Mutual relationship to the same rights of property or contract relationship. The doctrine of privity provides that a contract cannot confer rights or impose obligations arising under it on any person or agent except the parties to the contract.

Procurement The acquisition of goods or services.

Screening System Process to determine if a seller has the minimum qualifications to participate in the bidding process.

Severability   If any provision of the terms of use shall be deemed unlawful, void, or for any reason unenforceable, then that provision shall be deemed severable from the remaining terms of use and shall not affect the validity and enforceability of any remaining provisions.

Single Source   A seller / contractor is selected by a buyer to provide all procured goods and services

Sole Source   A seller is the only available source of a particular item that must be procured.

Waiver   Voluntary relinquishment or surrender of some right or privilege.

Warranty   A statement, either written, expressed, or implied, providing assurance that some specified provision in a contract is true. A protection plan against major repairs and breakdowns.

PMBOK® Guide—**Fourth edition, Project Procurement Management Processes**

The major processes of Project Procurement Management are:

Plan procurements – Documenting project procurement decisions. This process is concerned with determining what products or services must be acquired from outside the organization and what will be produced internally.

Conduct Procurements – The processes associated with obtaining seller responses, selecting sellers and awarding contracts

Administer Procurements – The processes associated with managing procurement relationships, monitorig contract performance, and making changes and corrections.

Close Procurements – The processes associated with terminating contracts,, finalizing open claims, updating records and closing completed procurements.

Activities, Questions, and Exercises

Refer to Chapter 19 of *Project Management: A Systems Approach to Planning, Scheduling, and Controlling* (10th Edition) for supporting information. Review each of the following questions or exercises and provide the answers in the space provided.

The following questions and exercises are associated with the knowledge areas of the *PMBOK® Guide*: Project Procurement Management and Project Risk Management.

PM Knowledge Note

The objective of the award cycle in contract negotiations is to reach agreement on a contract that will result in reasonable contractor risk and provide the contractor with the greatest incentive for efficient and effective economic performance.

1. **Make-or-Buy Decision.**

Review the list of factors that influence the make-or-buy decision and place them in the appropriate column.

The Make Decision	The Buy Decision

Factors:

   A. Less costly
   B. Easy integration of operations
   C. Utilize skills of suppliers
   D. Utilize existing capacity that may be idle
   E. Maintain direct control
   F. Small volume requirement
   G. Having limited capacity or capability
   H. Avoid unreliable supplier base
   I. Maintain design/production secrecy
   J. Augment existing labor force
   K. Maintain multiple qualified vendor sources
   L. Stabilize existing workforce
   M. Indirect control

2. **To Rent or Lease—That Is the Question**

The lease/rent decision is a financial endeavor that requires some analysis before making a decision. Consider the following situation:

You have the option to rent a piece of equipment for $200 per day. You can also lease the equipment for $70 per day plus a one-time cost of $8,000. You need the equipment for 75 days. Should you rent or lease?

   A. Rent _____ Lease ___☒___
   B. What is the break-even point in days? _____

> **PM Knowledge Note**
>
> *Before accepting the terms of a contract, many organizations conduct a detailed risk assessment using a third party that is not directly connected to the negotiations. This provides a more objective approach to the assessment and may identify significant risks that were not noticed during the negotiations process.*

3. To ensure that no contractor has more knowledge about a request for a proposal, the buyer schedules a _____ to provide an opportunity for all potential contractors to ask questions about the proposal and provide the same information to all interested parties.

a) Seller audit
b) Bidder's conference
c) Phase review
d) Request for bid

☑ **Study notes:**

The procurement process may include the use of several different documents:

- The Request for Information (RFI)—A document used to obtain information from potential providers before a formal request for proposal is issued. A document issued by a buyer to find providers or check that service levels of existing providers will meet the needs of the buyer.
- The Request for Quotation (RFQ)—A request for a final price quotation for a precise set of requirements.
- The Request for Proposal (RFP)—The formal mechanism or document by which a company conveys its business requirements to potential contractors/suppliers. An open invitation to potential suppliers to provide a proposal to meet the needs of the buyer.

Negotiation

The primary goal of negotiation should be to achieve a deal that both parties involved will agree to and that will accomplish the objectives established by both sides. Negotiation may require compromising, accommodating, and adjusting of demands before agreement can be reached. In some cases it will be necessary to identify common ground from which negotiations can begin.

Study notes: Tactics for negotiation:

- Ask open-ended questions—why? who? when?
- Probing—Ask questions that will help identify the other party's underlying needs. Example—I am not sure I understand why this point is so important to you. Can you explain your issues and concerns to help me understand?
- Control the documentation—Take notes or minutes to capture the information discussed. This will help keep the results of the negotiation accurate and fair.
- Develop alternatives (BATNA)—What other steps may be taken to satisfy specific needs? BATNA means Best Alternative to Negotiated Agreement. When the original negotiations toward an agreement are not successful, alternatives are introduced that may have a better chance of reaching agreement.

- Awareness of hygiene factors—Location selection for negotiations, type of table (round or rectangular), time of day, size of the room, room set-up, who sits facing the window.
- Set objectives for the minimum that will be accepted or the maximum that will be offered.
- Do not offer your minimum or maximum objectives at the start of negotiations. These are your objectives and will help to measure your success.

PM Knowledge Note

Negotiations should be planned for. The activities associated with negotiations include:
- *Developing objectives—what is the minimum that will be accepted? What is the maximum that will be offered?*
- *Evaluate the other party—strengths, weaknesses*
- *Define your strategy and tactics*
- *Gather all the facts you can*
- *Perform a complete price/cost analysis (Do your homework!)*
- *Arrange the "hygiene factors" to accommodate the negotiations*

4. ✍ **PM Quick Check:**

PMP® & CAPM® Exam

A. Before negotiations begin it is a good idea to determine what motivates the other party.

True_____ False_____

B. A major factor in the negotiations process is the willingness and ability to compromise.

True_____ False_____

C. The basic elements of a contract. Match the terms with the explanation:

1. Mutual Agreement		**A.** Must be a payment
2. Consideration		**B.** Disputes may be remedied in court
3. Contract Capability		**C.** Offer and acceptance
4. Legal Purpose		**D.** The agreement constitutes a recognized business or specifically stated reason
5. Form provided by law		**E.** The contractor has the ability to actually perform the required work

5. **Contract Forms**

Select the appropriate number for each situation.

A. When a contractor is required to deliver a definitive end product the contract is known as a

_____ contract.

B. If a contract specifies a level of effort that must be performed over a period of time and is based on skill sets and person hours the contract is known as a

_____ contract.

C. The final contract when agreed upon and signed is known as the _____ contract.

D. If work must begin before the contract has been through the complete approval and acceptance process to avoid late acquisition of resources or to avoid potential environmental problems a _____ contract may be approved to allow work to begin while the formal approval process continues.

 1. Definitive
 2. Term
 3. Completion
 4. Letter contract or letter of intent

PMP® &
CAPM®
Exam

6. Contract Types

A. Who has the greatest risk in a Firm Fixed Price or Lump Sum contract? _____

B. Who has the greatest risk in a Cost Plus type contract? _____

C. Who has the potential to benefit from a Cost Plus Incentive Fee contract? _____

D. In a Cost Plus Fixed Fee contract who pays for all project costs? _____

E. In the Cost Reimbursable type contracts who generally has the greatest risk? _____

F. The Cost Plus Percentage of Cost contract (CPPC) reduces the motivation of the _____ to complete the contract and may result in enormous cost to the _____

 1. Seller   **2.** Buyer

7. Calculating Project Costs

Review each of the following examples and determine the final price paid by the buyer. In some cases, a buyer and seller may negotiate an incentive type project that will benefit both sides if the conditions are met. Incentive contracts may be developed to speed up completion of a project or reduce buyer costs.

Example: Cost Plus Incentive Fee Contract

Target Cost	$100,000
Target Fee	$20,000
Target Price	$120,000
Sharing Ratio	70/30 (70% to buyer, 30% to seller)
Actual Cost	$90,000

Additional fee = $100,000 – $90,000 (target cost – actual cost) = $10,000 x .30 = $3,000
Total fee paid = $20,000 (agreed upon fee) plus $3,000 incentive = $23,000
Total project price = $23,000 + $90,000 = $113,000

A. Your Turn

Cost Plus Incentive Fee Contract

Target Cost $150,000

Target Fee	$25,000
Target Price	$175,000
Sharing Ratio	80/20 (80% to buyer, 20% to seller)
Actual Cost	$140,000

Additional fee =
Total fee paid =

Total project price =

B. Fixed Price Incentive Fee Contract

Target Cost	$150,000
Target Fee	$25,000
Target Price	$175,000
Sharing Ratio	60/40 (60% to buyer, 40% to seller)
Ceiling Price	$180,000
Actual Cost	$145,000

Additional fee =
Total fee paid =

Total project price =

C. Fixed Price Incentive Fee Contract

Target Cost	$150,000
Target Fee	$25,000
Target Price	$175,000
Sharing Ratio	60/40 (60% to buyer, 40% to seller)
Ceiling Price	$185,000
Actual Cost	$180,000

Additional fee =
Total fee paid =

Total project price =

8. More Contractions! Cost Plus Incentive Fee Contract (CPIF)

In the CPIF contract, the seller receives a bonus for performing the work below a target cost. As an example, consider the data below provided for a CPIF contract:

Target cost: $200,000
Target fee: $14,000
Sharing ratio: 80/20
Profit ceiling: $18,000
Profit floor: $4,000

In this example, the seller has agreed upon a target cost of $200,000 and a target fee of $14,000. If the seller performs the work at a lower cost than the $200,000, then the seller will receive additional profit up to a maximum profit of $18,000. If the seller performs the work for more than the target cost of $200,000, then the seller will receive less profit, but at least the minimum of $4,000.

The sharing ratio represents the amount of the overrun or underrun shared by each party. The first number represents the buyer and the second number represents the seller. If the work is performed below the target cost the buyer keeps 80% of the underrun and the seller keeps 20% of the underrun as extra profit up to a maximum profit of $18,000. If the contract is overrun, the buyer pays 80% of the overrun and the seller pays 20%, which is subtracted from the target profit as long as the profit is no less than $4,000. This is shown in the figure below:

Principles of Incentive Contracts

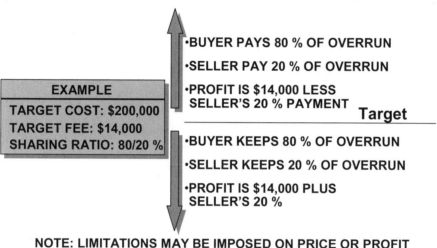

NOTE: LIMITATIONS MAY BE IMPOSED ON PRICE OR PROFIT

Consider the following example where the work is performed at $190,000.

Cost saving is $10,000.

Seller keeps 20% x 10,000 = $2,000 in extra profit.

Seller's total profit is therefore $16,000.

Seller will be reimbursed actual cost plus profit, which equals $206,000.

Now let's assume the seller spent $240,000 performing the work.

Cost overrun is $40,000.

Buyer pays 80% or $32,000.

Seller pays 20% of $40,000 or $8,000, which is subtracted from profits, resulting in a total profit of $6,000.

Seller will be reimbursed actual cost plus profit, or $246,000.

Now it's your turn. For each of the four situations below, and using the same data from the above problem, calculate the amount of money reimbursed to the seller if the seller's actual cost was:

a. $260,000
b. $230,000
c. $180,000
d. $140,000

9. An Incentive to Learn! Fixed Price Incentive Fee Contract (FPIF)

The fixed price incentive fee contract is very similar to the CPIF contract with the exception that instead of having the floor and ceiling on the profits, there is a ceiling on the final price paid by the seller. Using the data below:

Target cost: $200,000
Target fee: $14,000
Sharing ratio: 80/20
Price ceiling: $250,000

The price ceiling is the point of total assumption, or the maximum amount that the buyer will reimburse the seller. At the point of total assumption, the contract becomes a firm fixed price contract and the seller incurs all cost overrun from that point forth with no sharing.

Let's use the two examples from the above problem where the actual costs were $190,000 and $240,000. The problem progresses the same way and we calculate the contract price, which was $206,000 and $246,000, respectively.   In both cases, since the contract price did not exceed the price ceiling of $250,000, the seller will receive the full value of the price. If the price were $255,000, the seller would be reimbursed only $250,000.

It is possible that the actual cost can be so large that all of the profit will be lost. However, negative profit cannot exist. The most the seller can lose in profit is the target profit.

Now it's your turn. Assume that each of the four choices below is the seller's actual cost for performing the work. How much will the seller receive from the buyer at the completion of the contract?

a. $300,000
b. $260,000
c. $250,000
d. $240,000

10. Relative Contract Risk

Each type of contract has some degree of risk that is shared between the buyer and the seller. Consider the following nine contract types:

a) Cost Plus Incentive Fee (CPIF)
b) Cost Plus Award Fee (CPAF)
c) Firm Fixed Price (FFP)
d) Cost Plus Percentage of Cost (CPPC)
e) Cost Sharing (CS)
f) Cost (C)
g) Firm Fixed Price with Economic Adjustment (FFE)
h) Fixed Price Incentive Fee (FPIF)
i) Cost Plus Fixed Fee (CPFF)

In the figure below, label each of the nine triangles, which represent each contract type, as to how the risk is shared between the buyer and seller.

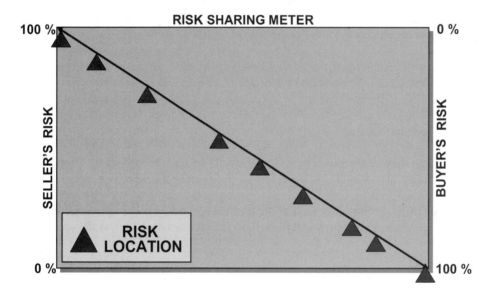

Relative Contract Risk

11. Procurement Options

There are four basic options available for procurement:

A. Procure all goods/services from a single source
B. Procure all goods/services from multiple sources
C. Procure only a small portion of the goods/services
D. Procure none of the goods/services

For each of the situations below, select the most likely option that would be selected:

1. You have the capability to produce a product. If the marketplace likes your product and demand increases sharply, the demand will most likely exceed your manufacturing capability. To minimize future risks, you decide to select one supplier as a backup and give them a small portion of the production contract. In this case, you would most likely be using option _____.

2. The demand for your product has exceeded expectations and you have decided to outsource all of the work. To minimize your dependency (i.e., risk) with using one contractor who might eventually "hold you hostage," you would most likely select option _____.

3. You have performed a make-buy analysis and discovered that it is more economical for you to manufacture the product yourself. The most likely option would be option _____.

4. Your company maintains a list of preferred suppliers all of whom could individually satisfy your production requirements. However, one supplier has continuously tried to keep costs down for the benefit of both you and the supplier. The most likely option would be option _____.

5. Your company had decided to manufacture all of the goods in-house. However, you have just won a large contract which may necessitate that the remaining portion of the production be outsourced so that manufacturing capability will exist for the new project. The most likely option selected would be option _____.

12. Make Versus Buy

During procurement planning, the project manager must decide whether to produce the goods/services internally to buy the goods/services externally. For each of the statements below, select whether the most likely option would be to make it or buy it:

1. Some suppliers can do it cheaper than you can.   Make: _____ Buy: _____

2. Management wants to stabilize the workforce.   Make: _____ Buy: _____

3. Your needs are one-of-a-kind and low volume.   Make: _____ Buy: _____

4. You prefer direct control over quality and cost.   Make: _____ Buy: _____

5. The product has proprietary knowledge in it.   Make: _____ Buy: _____

6. Your in-house capacity is limited.   Make: _____ Buy: _____

7. You want to augment the existing labor force.   Make: _____ Buy: _____

8. The preferred suppliers have superior skills.   Make: _____ Buy: _____

9. You want indirect control to minimize risks.   Make: _____ Buy: _____

10. The supplier base is unreliable.   Make: _____ Buy: _____

13. Rent Versus Lease

Your company has a need for some specialized equipment that can be leased or rented. If you rent the equipment, the cost is $100 per day. If you prefer to lease the equipment, which is usually for a longer period of time, the cost is $2,000 (nonrefundable payment) plus $50 per day. What is the breakeven point where renting and leasing are the same?

Let X = the breakeven number of days

Renting = $100X$

Leasing = $2,000 + $50X$

Therefore, $100X = $2,000 + $50X$

Solving for X, we find that X = 40 days. Therefore, if you plan on using the equipment for less than 40 days, the cost-effective decision is to rent. For more than 40 days of use, the correct decision is to lease.

Now it's your turn. A company needs some equipment for 120 days to meet contractual requirements. The rental cost is $200 per day and the leasing cost is $10,000 plus $75 per day. Should the company rent or lease?

14. Request Seller Response

    Below are several statements that affect procurement and contracting. Beside each statement, state whether or not this item would be part of the Request Seller response process.

    1. Name of the project manager.   Yes: _____ No: _____

    2. Bidder conferences (how often and when).   Yes: _____ No: _____

    3. Name of the contract administrator.   Yes: _____ No: _____

    4. How payments will be made to the seller.   Yes: _____ No: _____

    5. Specialized documentation (i.e., compliance with OSHA, EEO, EPA, etc.). Yes: _____ No: _____

    6. Proposal evaluation criteria.   Yes: _____ No: _____

    7. Listing of qualified bidders expected to respond.   Yes: _____ No: _____

    8. The process for change control.   Yes: _____ No: _____

    9. Adverting in a trade journal   Yes _____ No _____

    10. Qualified sellers list   Yes _____ No _____

15. **Acquisition Methods**

    In the left column are three methods for acquisitions. Select a response from the right column that matches the left column.

    1. Request for information (RFI)_____
    2. Request for quotation (RFQ)_____
    3. Request for proposal (RFP)_____

    A. Request for a formal response but usually just a cost or price without any supporting information.

    B. A request to see if the information is available, and if so, can it be obtained through licensing, leasing, joint venture, or some other way.

    C. Request for a formal response including full pricing data, labor justification, backup data, management capability, previous experience, etc.

16. Contractual Terminology

Listed below are several contractual terms. Following the contractual terms are definitions. Match the definitions to the contractual terms.

1. Arbitration: _____
2. Breach of contract: _____
3. Contract: _____
4. Executed contract: _____
5. Force majeure clause: _____
6. Good faith: _____
7. Infringement: _____
8. Liquidated damages: _____
9. Negligence: _____
10. Noncompete clause: _____

11. Nondisclosure clause: _____
12. Nonconformance: _____
13. Penalty clause: _____
14. Privity of contract: _____
15. Waiver: _____
16. Warranty: _____

A. The relationship that exists between the buyer and the seller
B. A violation of one's legally recognized right
C. The settling of a dispute
D. An intentional relinquishment of a legal right
E. Honesty and fair dealing between all parties
F. To violate or break a legal obligation
G. Restrictions on the disclosure of proprietary knowledge
H. A promise that the facts stated are true
I. An agreement that can be enforced in a court of law
J. Work that does not conform to specifications or requirements
K. Release from liability resulting from acts of God, wars, etc.
L. Reasonable damages resulting from a breach of contract
M. The failure to act in a reasonable manner
N. An agreement, in financial terms, for failure to perform
O. A completed contract
P. Restrictions on working for a competitor within a certain time

17. Characteristics of Contracts

In the left column are characteristics of certain types of contracts. In the right column are the names of the contracts. For each item in the left column, identify a contract type in the right column.

1. An almost bottomless pit of funds whereby the seller is reimbursed all costs as well as a profit for all money spent. _____

2. A sharing formula for costs above and below a target as well as a ceiling and floor on profits. _____

3. A profit fixed in dollars rather than percent. _____

4. A profit based upon how well the buyer likes the end result. _____

5. A contract which does not allow for profits and where the seller pays part of the cost. _____

6. A contract with a point of total assumption. _____

7. A contract where the maximum risk resides with the seller. _____

A. Firm Fixed Price (FFP)
B. Cost Plus Incentive Fee (CPIF)
C. Fixed Price Incentive Fee (FPIF)
D. Cost Sharing (CS)
E. Cost Plus Percentage of Cost (CPPC)
F. Cost Plus Fixed Fee (CPFF)
G. Cost Plus Award Fee (CPAF)

18. The Contract Administrator

For each of the items below, state whether or not this item is usually the responsibility of the contract administrator for the buyer.

1. Preparing a report card on the seller.   Yes: _____ No: _____

2. Determining a breach of contract.   Yes: _____ No: _____

3. Selecting the seller's project manager.   Yes: _____ No: _____

4. Selecting the buyer's project manager.   Yes: _____ No: _____

5. Approval of scope changes.   Yes: _____ No: _____

6. Interpretation of specifications.   Yes: _____ No: _____

7. Production surveillance.   Yes: _____ No: _____

8. Issuing of waivers.   Yes: _____ No: _____

9. Resolution of disputes.   Yes: _____ No: _____

10. Project termination.   Yes: _____ No: _____

11. Payment schedules.   Yes: _____ No: _____

12. Project closeout.   Yes: _____ No: _____

13. Validation of a bid protest.   Yes: _____ No: _____

19. True-False Exercise

1. Project procurement strategies (as opposed to corporate procurement strategies) most frequently allow for sole source procurement. True _____ False _____

2. Part of contract negotiations is to prepare a complete price/cost analysis. True _____ False _____

3. Verbal contracts, based upon mutual agreement, are enforceable in a court of law.
   True _____ False _____

4. Contracts are enforceable in a court of law even if the contract was not for a legal purpose.
   True _____ False _____

5. Term contracts can be completed without any deliverables being produced.
   True _____ False _____

6. Each contract type is a sharing of risk between the buyer and seller. True _____ False _____

7. The point of total assumption is the point where the buyer pays all additional costs over the target price. True _____ False _____

8. A letter contract or letter of intent allows the seller to begin work prior to the signing of the final contract and spend a specified amount of the final contract price. True _____ False _____

9. An order of precedence clause in a contract states which deliverables in the contract must be completed first. True _____ False _____

10. Backcharging is when the buyer re-opens a closed-out contract and pays the seller to perform additional work. True _____ False _____

20. Contract Selection Criteria

The buyer usually has criteria from which the final contract type is based. For each of the items below, state whether you agree that this item could be part of the selection criteria.

1. Overall degree of cost and schedule risk.   Yes: _____ No: _____
2. Type and complexity of requirements.   Yes: _____ No: _____
3. Extent of price competition.   Yes: _____ No: _____
4. Cost/price analysis (i.e., validity).   Yes: _____ No: _____
5. Urgency of requirements.   Yes: _____ No: _____
6. Performance period.   Yes: _____ No: _____
7. Contractor's responsibility and risk.   Yes: _____ No: _____
8. Contractor's ability to perform earned value measurement.   Yes: _____ No: _____
9. Amount of concurrent contracts that the seller is now working on. Yes: _____ No: _____
10. Extent of seller subcontracting required.   Yes: _____ No: _____

21. Explain the advantages and disadvantages of the decision to use a single source as a supplier.

Advantages	Disadvantages

Kerzner "Quick tips" for the Project Management Institute PMP® and CAPM® EXAM

The subjects in this chapter are most closely associated with the areas of the *PMBOK® Guide*: Project Procurement Management.

A contract is a legally binding and enforceable agreement.

A contract may be created either by verbal or written agreement or by implication by the conduct or actions of the parties involved.

Remember—The major elements of a contract are: offer, acceptance, consideration, capacity, legal form.

Contracts should be reviewed and validated to ensure that the conditions and objectives within the contract will meet the objectives of the organizations involved.

Contracts should be read thoroughly and carefully to make sure that the basic terms are clearly stated and fully understood.

There are several types of contracts. The Firm Fixed Price Contract places the greatest amount of risk on the seller. The Cost Plus Type contracts place the greatest amount of risk on the buyer.

Incentive contracts are negotiated to achieve mutually agreed upon benefits between the buyer and seller. Sharing ratios are established to calculate the actual incentives. Incentive Type contracts are designed to accelerate project completion or keep costs as low as possible.

Make-or-buy decisions should be carefully analyzed to obtain the greatest benefits regarding cost, control, and quality.

A letter of intent or letter contract may be used to begin work before the final contract terms and conditions have been officially approved and accepted.

A constructive change is a change that causes the contractor to perform work differently than initially required due to the action or inaction of the customer or other parties. An example of a constructive change would be late or unsuitable customer-furnished property.

Additional tips and practice items for the PMP® exam are included in each chapter and in the section of the workbook entitled *PMP Exam and PMBOK Guide® Review.*

Answers to Questions and Exercises

1.

The Make Decision	The Buy Decision
A	A
B	C
D	F
E	J
G	K
H	M
I	
L	

2. A—Lease, B—61.5 days

3. b

4. A. True B. True

   C.

Mutual Agreement	C	A. Must be a payment
Consideration	A	B. Disputes may be remedied in court
Contract Capability	E	C. Offer and acceptance
Legal Purpose	D	D. The agreement constitutes a recognized business or specifically stated reason
Form provided by law	B	E. The contractor has the ability to actually perform the required work

5. A—3, B—2 , C—1, D—4

6. A—Seller, B—Buyer, C—Seller and Buyer, D—Buyer, E—Buyer, F—Seller, and Buyer

7. A—Your Turn

Cost Plus Incentive Fee Contract

Target Cost	$150,000
Target Fee	$25,000
Target Price	$175,000
Sharing Ratio	80/20 (80% to buyer, 20% to seller)
Actual Cost	$140,000

Additional fee = $150,000 – $140,000 (target cost – actual cost) = $10,000 x .20 = $2,000
Total fee paid = $25,000 (agreed upon fee) plus 2,000 incentive = $27,000

Total project price $27,000 + $140,000 = $167,000

B—Fixed Price Incentive Fee Contract

Target Cost	$150,000
Target Fee	$25,000
Target Price	$175,000
Sharing Ratio	60/40 (60% to buyer, 40% to seller)
Ceiling Price	$180,000
Actual Cost	$145,000

Additional fee = $5,000 x .40 = $2,000
Total fee paid = $25,000 + $2,000 = $27,000

Total project price = $145,000 + $27,000 = $172,000

C—Fixed Price Incentive Fee Contract

Target Cost	$150,000
Target Fee	$25,000
Target Price	$175,000
Sharing Ratio	60/40 (60% to buyer, 40% to seller)
Ceiling Price	$185,000
Actual Cost	$180,000

Additional fee = $150,000 − $180,000 = −$30,000 x .40 = −$12,000
Total fee paid = $25,000 − $12,000 = $13,000

Total project price = $180,000 + $13,000 = $193,000, but the ceiling price is $185,000. The total project price paid by the buyer is $185,000.

8. Cost Plus Incentive Fee

a. $264,000
b. $238,000
c. $198,000
d. $158,000

9. Fixed Price Incentive Fee

a. $250,000
b. $250,000
c. $250,000
d. $246,000

10. Relative Contract Risk

Relative Contract Risk

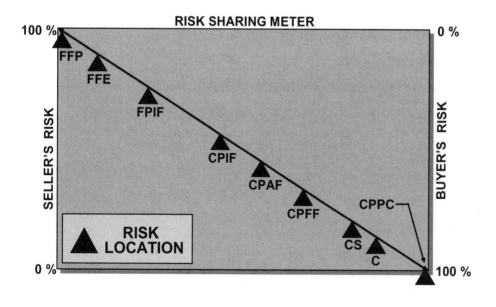

11. Procurement Options

   1. C (A is also possible)
   2. B
   3. D
   4. A
   5. A

12. Make Versus Buy

   1. Buy
   2. Make
   3. Buy
   4. Make
   5. Make
   6. Buy
   7. Make
   8. Buy
   9. Buy
   10. Make

13. Rent Versus Lease

The breakeven point is 80 days. Since the company needs the equipment for 120 days, the best choice is to lease.

14. The Solicitation Package

   1. No
   2. Yes
   3. No
   4. Yes
   5. Yes
   6. Yes
   7. Yes
   8. Yes
   9. Yes
   10. Yes

15. Acquisition Methods

   1. B
   2. A
   3. C

16. Contractual Terminology

   1. Arbitration: C
   2. Breach of contract: F
   3. Contract: I
   4. Executed contract: O
   5. Force majeure clause: K
   6. Good faith: E
   7. Infringement: B

    8. Liquidated damages: L
    9. Negligence: M
  10. Noncompete clause: P
  11. Nondisclosure clause: G
  12. Nonconformance: J
  13. Penalty clause: N
  14. Privity of contract: A
  15. Waiver: D
  16. Warranty: H

17. Characteristics of Contracts

  1. E
  2. B
  3. F
  4. G
  5. D
  6. C
  7. A

18. The Contract Administrator

  1. Yes
  2. Yes
  3. No
  4. No
  5. Yes
  6. Yes
  7. Yes
  8. Yes
  9. Yes
  10. Yes
  11. Yes
  12. Yes
  13. No

19. True-False Exercise

  1. True
  2. True
  3. True
  4. False
  5. True
  6. True
  7. False
  8. True
  9. False
  10. False

20. Contract Selection Criteria

All of the answers are possible selection criteria.

21. Explain the advantages and disadvantages of the decision to use a single source as a supplier. There may be several answers for advantages and disadvantages

Advantages	Disadvantages
Pricing advantages	No guarantee that pricing is competitive
Consistent level of quality	Potential for the seller / supplier to fail financially
Predicable results	Potential reduction in quality
	No back up supplier if the contracted supplier can not perform the work
	Potential delays in completing procured work

Your Personal Learning Library

Write down your thoughts, ideas, and observations about the material in the chapter that may assist you with your learning experience. Create action items and additional study plans to assist you in enhancing your skills or for preparing to take the PMP® or CAPM® exam.

Insights, key learning points, personal recommendations for additional study, areas for review, application to your work environment, items for further discussion with associates.

Personal Action Items:	
Action Item	**Target Date for Completion**

Chapter Twenty

QUALITY MANAGEMENT

The key to managing quality is to maintain a continuous awareness that quality is defined by the customer. Quality is a continuously improving process where lessons learned are used to enhance future products and services in order to retain existing customers, win back lost customers, and win new customers. The quality process includes planning for quality, organizing for quality, executing the work, monitoring the work, solving problems by identifying causes of poor quality, re-planning, and establishing preventive measures to reduce defects and poor quality.

There are several approaches and principles to achieving quality. These include:

- *Determining the cost of quality*
- *Zero defect programs*
- *Total Quality Management (TQM)*
- *Reliability engineering*

These principles and others have been established by organizations worldwide and are being implemented for the basic reason that quality is of strategic importance to a company. Quality is linked to profitability, it is a competitive weapon, it is an integral part of strategic planning, and it requires an organization-wide commitment. Every employee in any organization is responsible for quality.

PM Knowledge Note

Organizations generally agree that quality cannot be accurately defined by the organization. Quality is defined by the customer. Quality is associated with products and services that meet or exceed the need and expectations of the customer at a cost that represents outstanding value.

Glossary of terms Key terms and definitions to review and remember

Assignable Cause   A specific or special cause that requires investigation. Data points that are plotted outside of the upper and lower control limits of a control chart are reviewed to determine the assignable cause.

Attribute Sampling   A process in which the deliverable is assessed to determine if it either conforms or does not conform to specific acceptance criteria

Availability   The probability that the product, when used under given conditions, will perform satisfactorily when called upon.

Control Chart   A graphic technique for identifying whether an operation or process is in or out of control and tracking the performance of that operation or process against calculated control and warning limits. The dispersion of data points on the chart is used to determine whether the process is performing within prescribed limits and whether variations taking place (plotted and documented) are random or systematic (from an assignable or special cause).

Cost of Quality   The total cost incurred to prevent errors, appraise and inspect products, repair defects, and resolve non conformance issues. Cost of quality includes: the cost of conformance – prevention and inspection, and the cost of non-conformance – internal and external failure

Grade A category assigned to products or services having the same functional use but different technical characteristics. Example – a software application is considered high quality (no defects) but low grade (limited features and functions)

ISO 9000   A series of international standards that provides quality management guidance and identifies quality system elements. ISO 9001 is a voluntary international standard for quality management systems. The requirements for ISO 9001:2000 and ISO 9001:2008 are designed to be generic in nature so that they can be applied across all industries by both the public and private sectors.

Kaizen   Continuous incremental improvement of an activity to eliminate waste. Japanese process of continuous improvement using problem solving and analysis techniques. A quality improvement process that involves all managers and employees.

Maintainability   The ability of the product to be retained in or restored to a performance level when prescribed maintenance is performed.

Operability   The degree to which a product can be operated safely.

Pareto Diagram   A chart used to graphically summarize and display the relative importance of the differences between groups of data. A bar chart that displays frequency of cause of failure from left to right—greatest frequency to least frequent.

Produce-ability   The ability to produce the product with available technology and workers and at an acceptable cost.

Quality   An essential and distinguishing attribute of something or someone. A degree or grade of excellence or worth. The totality of features and characteristics of a product or service that bear on its ability to satisfy stated or implied needs.

Quality Policy   The overall intentions and direction of an organization regarding quality as formally expressed by top management. The general practical intention and direction of a company toward quality.

Reliability   The probability of the product performing without failure under given conditions and for a set period of time.

Rule of Seven   The rule of seven states that seven data points are required to indicate a trend or that seven data points are on one side of the mean.

Salability   The balance between quality and cost.

Social Acceptability   The degree of conflict between the product or process and the values of society.

TQM   Total Quality Management. A policy that includes all levels of employees in an organization and involves training, awareness, and overall acceptance of the quality mission and principles of the organization.

Variable Sampling   A process in which a product is measured on a continuous scale to determine the level of conformity.

Activities, Questions, and Exercises

Refer to Chapter 20 of *Project Management: A Systems Approach to Planning, Scheduling, and Controlling* (10th Edition) for supporting information. Review each of the following questions or exercises and provide the answers in the space provided.

The following questions and exercises are associated with the knowledge areas of the *PMBOK® Guide*: Project Quality Management.

PM Knowledge Note

Quality management is approached from many different perspectives. What works well in one organization may not work in another. It is important for an organization to establish a clear quality policy that can be communicated and instilled as part of the company culture.

1. The Shewhart Chart, adapted by W. Edwards Deming as a basis for managing quality, is associated with which of the following processes:

    a) Initiate Plan, Execute, Control, Close
    b) Plan, Do, Check, Act
    c) Prevention, Inspection, Internal failure, External failure
    d) Execute, Analyze, Respond, Monitor

2. Deming postulated that 85% of all quality problems required:

    a) Employee action
    b) Statistical analysis
    c) Management initiative and action
    d) The use of control charts

3. **Quality Guru Match Up**

    Match the principles with the appropriate quality guru.

    Deming _____

    Juran   _____

    Crosby _____

Taguchi _____

A. Trilogy of Quality Improvement, Quality Planning, and Quality Control
B. Design of experiments
C. Pioneered statistics and sampling
D. Fitness for use
E. Zero defects is possible
F. The legal implications of quality
G. Workers can't do their best. They must be shown what constitutes acceptable quality
H. Conformance to specifications
I. Quality is measured by the cost of nonconformance
J. Cost to correct problems rises the further the results are from target
K. Quality comes from prevention

4. Which of the following are true about Taguchi's approach to quality?

a) Quality should be designed into the product
b) Quality should be inspected into the product
c) Senior management must take a visible role in managing defects
d) The cost of quality is associated with nonconformance
e) Cost of quality is measured as a function of deviation from the standard

5. **Quality Control Charts**

Consider the data shown below which will be used to construct a quality control chart.

Data Point(N)	X	$(X-\bar{X})^2$
1	2.00	0
2	2.01	.0001
3	2.02	.0004
4	1.99	.0001
5	1.99	.0001
6	1.98	.0004
7	1.99	.0001
8	2.01	.0001
9	2.01	.0001
10	2.00	0
	20.00	0.0014

$$\bar{X} = 20.00/10 \text{ data points} = 2.00$$

$$\sigma = \sqrt{([X-\bar{X}]^2/[N-1])} = 0.01$$

Now let's apply this to a problem. Assume one of your customers wants this product and has given you a specification where the upper specification limit is at 2.05 and the lower specification limit is at 1.95. If you establish your process limits at X ± 3σ, then the upper control limit is 2.03 and the lower control limit is 1.97. This is shown in the figure below.

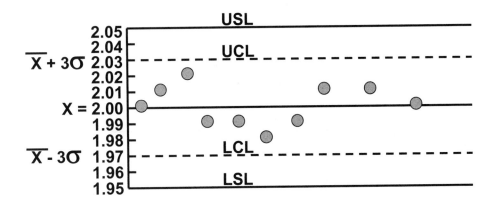

In this figure, the process is said to be "in control" because all of the data points fall within the upper and lower control limits. However, had the upper and lower specification limits been set to 2.015 and 1.985 respectively, the process would be "out of control" because of two of the points.

This example is for teaching purposes only. In most companies, 10 data points may not be a sufficient quantity to analyze the data in a control chart format.

Now it's your turn: Shown below are three control charts which represent special or assignable cause variability. For each of the control charts, select from the first group the reason why the process is out. Then, from the second group, select the cause or causes for the process to be out of control.

Group I: 1. A "Run"
                2. A point outside of the control limits
                3. "Hugging" the control limits

Group II: 1. Operator made a mistake
                2. Process may be out of alignment; shift in centerline
                3. Faulty raw material
                4. Operator pushed the wrong button
                5. Worn-out tooling

Control Chart A:

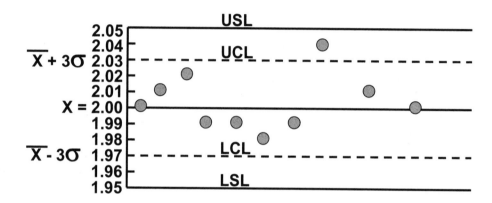

Control Chart B:

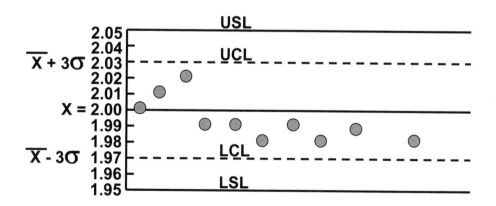

Control Chart C:

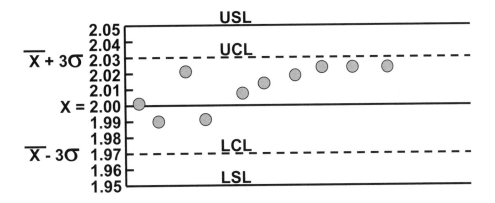

6. Quality Assurance and Quality Control

Review the list and determine which items are associated with Quality Assurance and which ones are associated with Quality Control.

Quality Assurance	Quality Control

A. Identify objectives and standards
B. Set standards that provide the basis for decisions regarding corrective action
C. Establish measurement methods
D. Multifunctional and prevention oriented
E. Plan for collection of data in a cycle of continuous improvement
F. Include scheduled reviews and audits
G. Comparison of actual results with standards
H. Take corrective action
I. Monitor and calibrate measuring devices

7. The Cost of Quality

The cost of quality is divided into two major sections: The cost of conformance and the cost of nonconformance. Match the list of costs with the appropriate cost of quality element.

Type of Cost	Cost of Quality Element
Internal Failure	
Prevention	
Appraisal	
External Failure	

A. Cost of conformance
B. Cost of nonconformance

8. Match the specific issue or event that may result in additional cost with the category of quality costs.

A. Prevention Costs   **B.** Appraisal Costs   **C.** Internal Failure   **D.** External Failure

Issue or Event	Quality Cost
Design review	
Rework	
Customer complaints	
Training	
Quality planning	
Scrap	
Process studies	
Repair	
Downtime	
Customer visits to inspect and take corrective action	
Inspection	
Lab tests	

9. More Guru Matching Exercises

1. In the left column are the names of three of the pioneers in project quality management. In the right column are their definitions of quality. Match up the two columns.

A. Deming _____   a. Conformance to requirements

B. Juran _____   b. Fitness for use

C. Crosby _____   c. Continuous improvement

2. Each of the pioneers in the left column had an approach they followed to improve quality. Match the approach to the pioneer.

A. Deming _____   a. The Trilogy

B. Juran _____   b. The Four Absolutes

C. Crosby _____   c. Plan-Do-Check-Act Cycle

3. Each of the pioneers in the left column was also known for some other contribution to the quality effort. Match the pioneer with the contribution.

A. Deming _____   a. Cost of Quality

B. Juran _____   b. The 85%–15% Rule

C. Crosby _____   c. Quality Is Free

10. True-False Exercise

1. Quality is inspection. True_____ False_____

2. Training to support total quality management begins with the executive levels of management.

   True_____ False_____

3. Quality is defined by the customer. True_____ False_____

4. Correction to quality problems should be documented. True_____ False_____

5. The alternative to Three Sigma is Six Sigma. True_____ False_____

6. Six Sigma applies to manufacturing only. True_____ False_____

7. Histograms are Quality Control tools. True_____ False_____

8. Quality circles are executive-level Quality Control teams. True_____ False_____

9. Pareto Charts are used to identify quality problems. True_____ False_____

10. Cause and Effect Charts are commonly used to identify the source of both quality problems and risks. True_____ False_____

11. If the Upper and Lower Control Limits are set at $X \pm 3\sigma$, then 99.73% of the products are expected to fall within these limits. True_____ False_____

12. Project managers have accountability for project success but can delegate to others the project manager's responsibility for quality. True_____ False_____

11. The following table is an example of a

    a) Pareto diagram
    b) Ishikawa diagram
    c) Control chart
    d) Check sheet

Defect	Supplier A	Supplier B	Supplier C	Supplier D	Total
Incorrect Invoice					
Incorrect Inventory					
Damaged Material					
Incorrect Test Documentation					
Total					

12. ✍ **PM Quick Check:**

Which of the following charts is most commonly associated with cause-and-effect analysis?

a) Check sheet
b) Pareto diagram
c) Histogram
d) Fishbone or Ishikawa diagram

PM Knowledge Note

Pareto diagrams are associated with the 80/20 rule. This rule states that 80% of problems can be associated with 20% of the identified causes.

13. ✍ **PM Quick Check:** True or False?

PMP® & CAPM® Exam

A. The use of a control chart is focused on prevention of defects.

True _____   False _____

B. There are two types of control charts—variable charts and attribute charts.

True _____   False _____

C. Attribute sampling is focused on a pass/fail or conform/does not conform decision basis.

True _____   False _____

D. Six Sigma is a method of quality management that focuses on the control of a process to achieve 3.4 defects per million opportunities of failure.

True _____   False _____

E. The project sponsor is ultimately responsible for the quality of the project results.

True _____   False _____

F. Total Quality Management is a program that is targeted specifically for the upper management of an organization.

True _____   False _____

G. Match the terms:

A. Mean   1. Most frequent
B. Mode   2. Middle value
C. Median   3. Average

14. What is meant by the 80/20 rule?

15. Provide examples of the cost of quality component – Internal failure

16. Explain the difference between accuracy and precision.

☑ **Quality Study Notes:** Quality management is often about statistics and analysis. It is important to become familiar with the terminology of statistical analysis.

- Population—the universe of environment that will be analyzed
- Variable sampling—a process in which products are evaluated on a continuous scale that allows for some deviation from the standard as long as it falls within set parameters
- Attribute sampling—a process that determines acceptance or rejection of a product. There is no flexibility and decisions are either pass/fail or conform/does not conform
- Producer's Risk—known as Alpha risk. The risk to a producer that a good lot will be rejected
- Consumer's Risk—known as Beta risk. The risk that a consumer may accept a bad lot
- Normal Distribution—associated with the bell-shaped curve, it represents a theoretical frequency distribution of measurements. The normal distribution curve, according to statistics, indicates that there is a probability of 68.26% that the results will fall within one standard deviation from the mean, 95.46% within two standard deviations and 99.73% within three standard deviations from the mean. The normal distribution curve provides a probability confidence factor.

Kerzner "Quick-tips" for the Project Management Institute PMP® and CAPM® EXAM

The subjects in this chapter are most closely associated with the areas of the *PMBOK® Guide*: Project Quality Management.

Project Quality Management includes three major processes as described by the *PMBOK® Guide*–Fourth edition

Plan Quality – Defining quality requirements and then documenting how the project will achieve compliance

Perform Quality Assurance – Auditing quality requirements and results from quality control measurements to ensure appropriate levels of quality are achieved and standards are in use

Perform Quality Control –The process of monitoring and recording results of executing quality activities to assess performance and recommend necessary changes.

Remember that the customer defines quality.

The project manager has ultimate responsibility for quality in the project environment.

Prevention over inspection is the main goal of a quality process.

Cost of Quality includes the Cost of Conformance and the Cost of Nonconformance.

> The cost of conformance includes prevention and appraisal

> The cost of non conformance includes internal and external failure

Quality policies and procedures will vary greatly from one company to another. A general awareness of quality principles and tools and techniques is important for the PMP® exam.

Remember the differences between Quality Assurance (process) and Quality Control (action based on measurements).

Quality is a continuous process of improvement.

The Shewhart Chart, adapted by Deming, displays the process of Plan, Do, Check, Act.

Crosby believes that zero defects can be achieved and that the cost of quality is associated with non conformance

Juran was focused on Fitness for use, conformance to specifications, safety, and the legal aspects of quality – warranties, liabilities

Taguchi developed the Design of Experiments process. He believed that quality should be designed in, not inspected in. As results move further from target, it becomes more costly to return to the target.

Control charts will assist in the identification of assignable causes. The main components of a control chart are the mean or process average, the upper control limit (UCL and lower control limit (LCL). Control charts will identify trends, runs, processes in control, and processes out off control.

Remember the various Quality Control tools—Control charts, Pareto diagrams, check sheets, cause-and-effect diagrams, scatter diagrams, attribute sampling, variable sampling.

Additional tips and practice items for the PMP® exam are included in each chapter and in the section of the workbook entitled *PMP® Exam and PMBOK Guide® Review*.

Answers to Questions and Exercises

1. a

2. c

3. Deming   C,G,J
   Juran   A,D,F
   Crosby   E,H,I,K
   Taguchi B

4. a

5. **A.** I—2, II—1,3,4
   B. I—1, II—2
   C. I—3, II—5

6.

Quality Assurance	Quality Control
A, D, E, F	B, C, G, H, I

7.

Type of Cost	Cost of Quality Element
Internal Failure	B
Prevention	A
Appraisal	A
External Failure	B

8.

Issue or Event	Quality Cost
Design review	A
Rework	C
Customer complaints	D
Training	A
Quality planning	A
Scrap	C
Process studies	A
Repair	C
Downtime	C
Customer visits to inspect and take corrective action	D
Inspection	B
Lab tests	B

9.

   1. A. c
     B. b
     C. a

   2. A. c
     B. a
     C. b

   3. A. b
     B. a
     C. c

10.

   1. False
   2. True
   3. True
   4. True
   5. True
   6. False
   7. True
   8. False
   9. False
   10. True
   11. True
   12. False

11. d

12. d

13. A True
   B. False
   C. True
   D. True
   E. False
   F. False

14. What is meant by the 80/20 rule? 80% of problems are associated with 20% of the causes

15. Provide examples of the cost of quality component – Internal failure
   Rework, scrape, defects, repairs

16. Explain the difference between accuracy and precision.

Accuracy is defined as, "The ability of a measurement to match the actual or real value of the quantity being measured".
Precision is defined as, "(1) The ability of a measurement to be consistently reproduced" and "(2) The number of significant digits to which a value has been reliably measured".

Your Personal Learning Library

Write down your thoughts, ideas, and observations about the material in the chapter that may assist you with your learning experience. Create action items and additional study plans to assist you in enhancing your skills or for preparing to take the PMP® or CAPM® exam.

Insights, key learning points, personal recommendations for additional study, areas for review, application to your work environment, items for further discussion with associates.

_____

_____

_____

_____

_____

_____

_____

_____

_____

_____

Personal Action Items:	
Action Item	**Target Date for Completion**

Chapter Twenty-One

MODERN DEVELOPMENTS IN PROJECT MANAGEMENT

Project management has been in existence for thousands of years. From the time the pyramids were constructed, probably even well before then, there were projects and project managers. Over the centuries project management has changed and improved as new tools and techniques were developed to gain efficiencies in the use of resources and the management of time. In the past few decades, there has been an increased interest in project management not only at the single project level but also at the program and portfolio level. Organizations are now interested in how to further financial gains, increase productivity, and reduce costs through organizational or enterprisewide project management.

In the early to mid-1990s, the project management office (PMO) became a focus of attention as organizations saw value in standardizing their project management processes. The capability maturity model (CMM) also became an item of interest to many corporate executives. Continuous improvement through Six Sigma, the development of competency models, and development of methods for managing multiple projects soon followed.

Today, there is continued interest in each of these areas including a new edition introduced by the Project Management Institute (PMI®) in 1999—OPM3®, or Organizational Project Management Maturity Model. OPM3® introduces a standard for measuring project management maturity in an organization through assessing the use of generally accepted best practices in project management. Other methodologies and techniques are also being used by many companies. These techniques include Agile Project Management, Lean Management, and Value Driven Project Management

The following are the major areas of interest that have surfaced as project management continues to develop:

- The Project Management Maturity Model (PMMM)
- Developing effective procedural documentation (standards and guidelines)
- Project Management Methodologies (at the project, program, and portfolio levels)
- Continuous Improvement (CMM and Six Sigma)
- Capacity Planning
- Competency Models (organizational and project management)
- Management of Multiple Projects
- End of Phase Review Meetings (Project Evaluations and Reviews)

- Critical Chain or Theory of Constraints
- OPM3 – Organizational Project Management Maturity Model
- Rational Unified Process – RUP
- Agile Project Management

The Project Management Maturity Model (PMMM)

The goal of PMMM is to assist an organization in achieving excellence in project management through a five-step process.

Level 1—Establish a common language. The first step is to ensure a basic knowledge of project management and an understanding of the accompanying terminology and language of project management.

Level 2—Common Processes. Processes that are used repeatedly on projects that are successfully completed, are defined and developed and then applied to other projects to increase the probability of success. These processes are recognized as also contributing to or supporting other processes within an organization.

Level 3—Singular Methodology. Recognition of the synergistic value of combining organizational processes into a singular methodology that improves the ability to control project outcomes and promote consistency within an organization.

Level 4—Benchmarking. Recognizing that process improvement is necessary to maintain a competitive advantage. Comparison of project results to organizational goals and industry best practices is necessary and must be performed on a regular basis.

Level 5—Continuous Improvement. The evaluation of information obtained through benchmarking and the actions taken to improve a process.

Although depicted in the model as a series of steps, these levels and their associated activities may overlap depending upon the capabilities of the organization.

The Project Management Maturity Model can be expected to impact the culture of an organization as processes are developed and longstanding procedures and positions of authority are affected, and in many cases, change considerably as more formal project management processes are introduced.

Developing Procedural Documentation

Procedural documentation will increase efficiencies, establish more consistency in planning projects, and assist an organization in achieving project management maturity.

Benefits of procedural documentation:

- Establishment of guidelines and greater uniformity
- Less but more useful documentation
- More effective and clear communication
- Standardized reports, forms, and data
- Greater unity within and across project teams
- Basis for analysis
- Minimize paperwork
- Reduction in conflict over procedures

Project Management Methodologies (at the project, program, and portfolio levels)

Establishing methodologies for managing projects, programs, and portfolios will increase the likelihood of success at each of the levels. This greatly depends upon the acceptance of these methodologies by the project managers and project teams assigned to projects and programs. Care should be taken to introduce methodologies with a focus on the impact to organizational culture. It may take significant time for methodologies to take hold, and if not properly introduced, they could meet with great resistance from project personnel. Project Management Offices can be used to facilitate the introduction of methodologies and changes to organizational culture.

Continuous Improvement (CMM and Six Sigma)

Continuous improvement can be associated with the Deming/Shewhart Model—PDCA or Plan, Do, Check, Act. CMM, or Capability Maturity Model, is very similar to the PMMM model and encourages an organization to progress through levels to achieve maturity. Six Sigma is considered a strategic approach to continuous improvement through tighter and tighter quality controls designed to reduce or possibly eliminate product defects. Six Sigma focuses on the acronym DMAIC – Define Measure Analyze, Improve, Control.

Capacity Planning

The improvements gained through effective project management are measured through increased productivity, the need for fewer resources, and completing more work in less time. Basically, through project management, an organization can increase its "capacity" to perform work. The gains obtained through efficient processes and methodologies impacts an organization's business forecasts by requiring adjustments in items such as anticipated growth, future staffing, and new product development.

Competency Models

Job descriptions were used to provide information about what an employee was expected to do. They did not address the level of competency and mainly focused on the expectations of management. In today's project environment, project managers are expected to assume many roles including leader, planner, and manager. Competency models will be used to determine the skill levels and ability of project managers which, in turn, will be used to identify where additional skills training and development may be needed. Competency models will address and score abilities in such areas as project management (planning, scope definition, WBS development), interpersonal skills (conflict management, motivation, communication), business acumen (organizational processes, financial management), and leadership.

Management of Multiple Projects

Today's project environment often requires a project manager to manage several projects concurrently. These projects may range from very small projects with short durations to more complex projects with longer durations. The span of control (number of projects assigned) to a project manager is usually dependent on previous experience, track record, and overall competency. Managing multiple projects presents new challenges to the project manager and a change from a single project mentality.

Areas where change will be required:

- Project selection—ensuring the right projects have been assigned
- Prioritization of projects—for effective resource management
- Management of scope changes—connections between projects
- Capacity planning
- Program and/or portfolio methods and procedures—new guidelines that allow a greater degree of flexibility in decision making

- Organizational structures—changing the existing structure to increase the effectiveness of the project management methodology. More influence and authority granted to the project manager

End of Phase Review Meetings (Project Evaluations and Reviews)

Establish processes for reviewing projects throughout the life cycle, not just at the end of the project. This is essential to ensure that the organization continues to support projects that will yield the greatest benefits to the organization. Establishing criteria for success by phase will increase the probability of successful completion of the project and assist an organization in managing its resources more effectively.

Critical Chain or Theory of Constraints

This is a method of planning and managing projects that puts more emphasis on the resources required to execute project tasks. It was developed by Eliyahu M. Goldratt. This method contrasts the more traditional Critical Path methods, which emphasizes task order and rigid scheduling. A Critical Chain project network will tend to keep the resources load leveled but will require them to be flexible in their start times and to quickly switch between tasks and task chains to keep the whole project on schedule. Project buffers and activities buffers are used to manage project work throughput. There is also emphasis on correcting and strengthening any weak links (resource constraints) that may be identified.

Organizational Project Management Maturity Model - Organizational Project Management Maturity Model — OPM3®

A standard developed by the Project Management Institute for reviewing, managing and enhancing organization project management through best-practices. OPM3® provides a way for organizations to understand their project management processes and measure their capabilities in preparation for improvement. OPM3® includes three major elements:

- **Knowledge -** project management best practices in use across many industries
- **Assessment** – Evaluation of current project management capabilities to identify areas in need of improvement.
- **Improvement -** Develop a plan based on the results of the assessment to achieve improvement goals.

Rational Unified Process - RUP

A prescriptive, well-defined system development process, often used to develop systems based on object and/or component-based technologies. It is based on planning and engineering principles such as taking an iterative, requirements-driven, architecture specific approach to software development. It focuses on relatively short-term iterations with well-defined goals and go/no-go decision points at the end of each phase, to provide greater visibility into the development process.

Agile Project Management – A highly iterative and incremental process involving developers and project stakeholders actively working together to produce a series of short gains where needs and functionality are identified and prioritized. It is a rapid planning process that involves several iterations, reviews and lessons learned and provides and opportunity for immediate feedback from stakeholders.

Activities, Questions, and Exercises

Refer to Chapter 21 of *Project Management: A Systems Approach to Planning, Scheduling, and Controlling* (10th Edition) for supporting information. Review each of the following questions or exercises and provide the answers in the space provided.

1. The five levels of the Project Management Maturity Model Are:,

2. How does Agile project management differ from what is known as " traditional" project management?

3. What does the acronym DMAIC refer to?

4. What is the major difference between the CPM (Critical Path Method) and Critical Chain Method?

5. Why is it important to identify and document best practices?. Why is continuous improvement a major factor in achieving project management maturity?

6. Why does an organization experience high levels of risk and difficulty when attempting to achieve level three of the project management maturity model?

7. What is the difference between a job description and a competency model?

_____
_____
_____
_____

Answers to Questions and Exercises

1. The five levels of the Project Management Maturity Model Are: Establish a common language, define common processes, create a singular methodology, benchmark, continuous improvement.

2. How does Agile project management differ from what is known as "traditional" project management? Traditional project management emphasizes a disciplined approach to planning in which several phases exist, many controls are in place, and progress is measured in terms of the entire project. Agile is an accelerated and iterative approach using short incremental steps that are analyzed for lessons learned.

3. What does the acronym DMAIC refer to? Define Measure Analyze, Improve, Control – It is associated with The SIX SIGMA approach to improving quality and reducing the number of defiects.

4. What is the major difference between the CPM (Critical Path Method) and Critical Chain Method?

CPM focuses on sequence of activities and activity duration to produce a critical path – the longest path through a project network. Critical chain method focuses on resource constraints and the use of project and activities buffers to relieve constraints.

5. Why is it important to identify and document best practices?. Why is continuous improvement a major factor in achieving project management maturity? Documenting best practices will improve overall project and organizational performance by making useful information available to multiple project teams that may reduce cost, accelerate project completion time frame, improve quality and reduce risk. Continuous improvement is the highest level of the project management maturity model and indicates that the organization has achieved a success based culture through a practice of analyzing performance to generate a steadily position of superiority over competitors.

6. Why does an organization experience high levels of risk and difficulty when attempting to achieve level three of the project management maturity model? The changes involved in establishing a methodology may have an effect on corporate culture, there may be resistance to changes in process, and an unwillingness to accept changes in authority and the decision making process.

7. What is the difference between a job description and a competency model? Job descriptions emphasize deliverables and specific expectation. Competency models emphasize skills required to achieve deliverables.

Your Personal Learning Library

Write down your thoughts, ideas, and observations about the material in the chapter that may assist you with your learning experience. Create action items and additional study plans to assist you in enhancing your skills or for preparing to take the PMP® or CAPM® exam.

Insights, key learning points, personal recommendations for additional study, areas for review, application to your work environment, items for further discussion with associates.

_____

_____

_____

_____

_____

_____

_____

_____

_____

_____

_____

Personal Action Items:	
Action Item	**Target Date for Completion**

Chapter Twenty-Two

SITUATIONAL EXERCISES

Managing actual projects is generally more difficult and unpredictable than what is described in the many books, articles, and seminars about project management that are available. Following processes that have been carefully explained in books is helpful in getting the project started and providing a guide to keep the project on track, but the real challenge is to use the information provided in the reference literature and apply the knowledge to the project. Project management is about the application of tools and techniques to a project to achieve success, and the project manager must become familiar with these tools and be able to determine which ones can and should be used during the project life cycle and which ones do not apply to the project. Good judgment is a key factor in managing projects, and an effective method for fine-tuning and enhancing your judgment is through situational analysis.

The exercises in this chapter will assist you in further developing your project management skills and will also help you if you are planning to apply for the PMP® credential. The PMP® exam tests the applicant on the principles of project management through situational-type questions. These questions place you, the student or applicant, in a project situation, not unlike something you may have experienced in an actual project, and you are expected to select the correct answer or solution to the problem that has been described. Situational exercises allow you to analyze a problem; consider the possible alternatives that may exist; factor in the risks, both negative and positive; reason through the problem; and determine the best solution. It is important to remember that there may be many alternatives, but there is generally one that would be considered the best choice when all factors have been considered.

THE STRUGGLE WITH IMPLEMENTATION

A small division (650 employees) of one of the big three automakers recognized the necessity for becoming more project-driven. As a result, a committee was formed to recommend the methodology and timetable to redefine the culture of the company.

The division had tried to use project management previously and found few successes. Cultural obstacles had to be overcome. The committee identified three major issues that had to be considered:

A. The only people allowed to function as project managers were the line managers, who were instructed to get the projects completed while managing their own lines.

B. Line employees were reluctant to use software for project tracking for fear that the software would identify the "truth" in their estimates.

C. Executive management was always too busy to meet with consultants who were brought in to discuss the implementation of change.

An organizational audit was sponsored by the committee. The audit was designed to identify problems that exist currently within the project teams. The results are shown in Exhibit 1.

1. How mature in project management is the organization?

_____

_____

_____

_____

2. Is there decentralization and delegation of authority? Explain.

_____

_____

_____

_____

_____

3. Can the problems in Exhibit 1 be classified into groups? If so, what are the groups?

_____

_____

_____

_____

_____

_____

4. Is there "visible" executive support for project management?

_____
_____
_____
_____
_____
_____

5. What recommendations would you make to the steering committee? To senior management?

_____
_____
_____
_____
_____
_____

6. How long do you think it will take for changeover to occur?

_____
_____
_____
_____
_____
_____

7. How successful will the changeover be? Explain your answer.

_____
_____
_____
_____
_____
_____

EXHIBIT 1

1. Continuous process changes due to new technology

2. Not enough time allocated for effort

3. Too much outside interference (meetings, delays, etc.)

4. Schedules laid out based on assumptions that eventually change during execution of the project

5. Imbalance of workforce

6. Differing objectives among groups

7. Using a process that allows for no flexibility to "freelance"

8. Inability to openly discuss issues without some people taking technical criticism as personal criticism

9. Lack of quality planning, scheduling, and progress tracking

10. No resource tracking

11. Inheriting someone else's project and finding little or no supporting documentation

12. Dealing with contract or agency management

13. Changing or expanding project expectations

14. Constantly changing deadlines

15. Last minute requirement changes

16. People on projects having hidden agendas

17. Scope of the project is unclear right from the beginning

18. Dependence on resources without having control over them

19. Finger-pointing: "It's not my problem"

THE BAD APPLE

John Doyle is the project manager for the Prism Project, a high-priority project that has also been assigned a project sponsor, from the executive levels. Two of Mr. Doyle's employees have been performing at levels substantially below expectation. These two employees report directly to a line manager, but are assigned full-time to the project. At a meeting with his director and immediate supervisor, Carol Brody, John expresses his frustration.

Ms. Brody: "Every project, sooner or later, will find team members that are bad apples. It was my understanding that good project managers should plan for this possibility."

Doyle: "Our manual on project guidelines does not discuss this at all, and I'm not sure that we can put all possible situations into our guide even if we want to."

Ms. Brody: "Whether we handle these situations formally (i.e., through guidelines) or informally is irrelevant to me. My concern is that we have some methodology to cover this. What did you attempt the last time this happened?"

Doyle: "Once I had a poor employee removed from the project. His close contemporaries working on the project lost enthusiasm for the project for fear that they would be next to be asked to leave. The project really suffered from poor morale. Another time I tried to give an employee an on-the-spot written performance review, and she laughed at me, telling me that I did not have the authority to do this. On a third occasion, I reassigned the employee to a less critical activity and this was partially successful. Unfortunately, I could not find a qualified replacement that late in the project to complete the vacated task."

Ms. Brody: "Let's make sure you understand that this problem is yours and should be resolved at the working levels. I really think there may be other options for you to consider at this time."

QUESTIONS

1. What other options are available?

    _____

    _____

    _____

    _____

SCORING SHEET

Each response you circled in Questions 1–20 had a column value between –3 and +3. In the appropriate spaces below, place the circled value (between –3 and +3) beside each question.

Embryonic

#1 _____

#3 _____

#14 _____

#17 _____

Total

Executive

#5 _____

#10 _____

#13 _____

#20 _____

Total

Line Management

#7 _____

#9 _____

#12 _____

#19 _____

Total

Growth

#4 _____

#6 _____

#8 _____

#11 _____

Total

Maturity

#2 _____

#15 _____

#16 _____

#18 _____

Total

Transpose your total score in each category to the table below by placing an "X" in the appropriate area.

Points													
Stages	−12	−10	−8	−6	−4	−2	0	+2	+4	+6	+8	+10	+12
Maturity													
Growth													
Line Management													
Executive													
Embryonic													

2. Can a project manager remove employees from a project if the employees report administratively to a line manager?

_____

_____

_____

_____

_____

3. Should the project sponsor be involved in this problem? If so, what help would you expect from the project sponsor?

_____

_____

_____

_____

THE COMMUNICATION PROBLEM

In each of the situations below, something appears to have gone wrong with the communication process. Identify where in the communication process the breakdown occurred (i.e., the message, encoding, decoding, feedback, etc.).

1. During a project team meeting, you inform your laboratory technician that the customer wants their product tested at 80°, 90°, and 100°F rather than at the normal testing temperatures of 75°, 85°, and 95°F. The tests are scheduled to begin two months from now. After testing is completed, you read the report prepared by the lab personnel and find that testing was performed at 75°, 85°, and 95°F.

2. Mary is the project engineer assigned to your project team. During a team meeting, Mary casually informs you that she will have her yearly performance review with her line manager next month, at which time she intends to request a promotion from "engineer" to "senior engineer." You wish her luck and take no further action.

   In the following month, you find that Mary's request for promotion was denied. The grapevine has it that you sabotaged her chance for promotion when, in fact, you never even talked to her line manager.

3. You are managing a project team of twenty people. Brenda is a procurement specialist who has never worked for you before. During a one-on-one meeting, you inform Brenda that you will need 2,000 pounds of an expensive raw material and you instruct her to find out as quickly as possible the lowest possible cost.

   Two weeks later, you receive a phone call from the inventory control group that your 2,000 pounds have arrived and have been placed into inventory stores. This creates a serious problem because the project's budget did not plan for early procurement or even 100 percent procurement at one time. You find out that Brenda signed the purchase requisition form for you.

4. The vice president for information systems informs you that he would like to be informed as to the status of your project and that he wants to meet with you a week from Tuesday. As you walk into his office, he says, "Where's the status report I asked you for?"

5. The Drafting Department has just assigned a draftsman to your project. You meet with the new team member, a person whom you have never met previously. You inform him of the way you want the drawing and ask him if he understood what you said. He responds, "Yes!"

   Two weeks later you receive the drawing and find out that it is not to your specifications. You question the draftsman, and he responds, "This is the standard way we always prepare these drawings."

6. You instruct one of your team members to prepare a technical report for next month's meeting with the customer. To prevent any communication gap from occurring, you provide the team member with an exact list of the technical details to go into the report. The week before the meeting, your team member wants you to review the report prior to reproduction. What you thought would be a thirty-five page report is now two hundred pages with multicolored artwork.

MEETINGS, MEETINGS, AND MEETINGS

In February 1994, Carver Company purchased a licensing agreement to place a "Meeting Scheduler" software package on the computers of about 400 employees. The organization became meeting-happy overnight. Any employee could look at the calendar of all other employees and schedule a meeting if the time slot were vacant. Some people would show up at work and discover that they were scheduled for six hours of meetings in one day without having agreed to attend. Others were scheduled for meetings where they had no idea what the meeting was about.

Carver's senior management realized quickly that the system was being abused. However, they were reluctant to remove that software because of its inherent value if used correctly. Another alternative had to be developed.

Carver's management instructed the Human Resources Group to prepare a meeting appraisal form. Every employee in every meeting would have to fill out this form. The completed forms would then be returned to the Human Resources Group for analysis. The results would be summarized in a memo, with copies to the individual requesting the meeting as well as to his/her superior. Copies would also be sent to all attendees.

The Human Resources Group prepared a list of topics to be included in the form. The form had to be a single sheet of paper, preferably on one side. The following information was considered to be critical:

A. Could the topics of this meeting have been covered without this meeting (i.e., memo, phone call, phone mail, one-on-one, etc.)?

B. Was an agenda sent out in advance listing the purpose of the meeting and what subjects would be covered and during which time periods?

C. Did the meeting start and finish on time?

D. Were the right people in attendance? Did they all show up?

E. Was the agenda adhered to?

F. Were there summary agreements made and possible action items?

G. Were the "meeting robbers" present that prevented the meeting from being conducted properly (i.e., the show-off, the abuser, the rambler, the assignment misser, the whisperers, the supersalesman, the wallflower, the later arriver, the wanna-be leader, etc.)?

H. How was the meeting rated overall (i.e., productive, partially productive, waste of time, etc.)?

QUESTIONS

1. Is the above list complete, or should there be additional topics included?

_____

_____

_____

_____

_____

_____

2. Using either one or two sheets of paper, design the meeting evaluation form. Make sure that you include optional "degrees" of answers, such as very well done, well done, acceptable, etc.

_____

_____

_____

_____

_____

_____

3. Does this form apply also to project team meetings? Explain your answer.

_____

_____

_____

_____

_____

THE EMPOWERMENT PROBLEM

In seminars to TQM, employees were told that empowerment must be provided to the quality teams. The process of empowerment worked reasonably well on the TQM teams, but struggled on project teams. The question, of course, was why. Some of the project teams were large enough to justify their own staff on a full-time basis. Other projects were so small that employees found themselves working on several projects at the same time.

QUESTIONS

1. Should the project team be empowered to set goals or should goals be set elsewhere?

2. Should a project team be empowered to make decisions at the end-of-phase reviews (i.e., open gates to the next phase)?

3. How does a project manager know whether his/her team is empowered to make decisions?

4. How does a project manager know whether employees are empowered to make decisions for their line managers?

5. How does a team resolve a problem if the team members are not empowered to make decisions?

_____

_____

_____

_____

_____

_____

6. Who empowers project team members?

_____

_____

_____

_____

_____

_____

7. Can empowerment work if project team members are part-time on several projects at once?

_____

_____

_____

_____

_____

_____

PROJECT MANAGEMENT PSYCHOLOGY

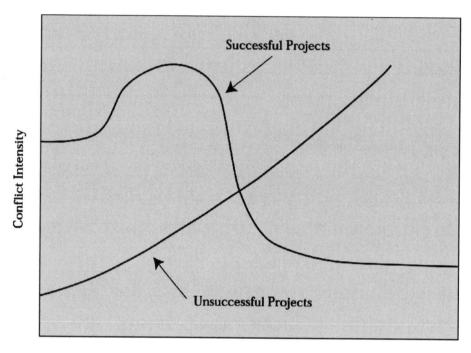

Time/Life-Cycle Phases

1. The figure shown above reflects current thinking concerning conflicts. Explain this figure.

_____

_____

_____

_____

_____

_____

2. Is it possible that the "bulge" may occur further to the right?

_____

_____

_____

_____

_____

_____

3. Can the bulge be pushed to the left, i.e., right to project initiation? If so, is there a psychology to project staffing? Explain your answer.

_____

_____

_____

_____

_____

_____

THE POTENTIAL PROBLEM AUDIT

There exist both subtle and not-so-subtle problems on projects that eventually lead to schedule slippages and cost overruns. The problem areas are both quantitative and nonquantitative. Several of these problems can occur on any given project. Exhibit 1 provides you with a list of 30 situations that lead to schedule slippages and cost overruns.

You and your team should perform the following:

1. Select a company that is known to use project management. (You may need the instructor's help.) The more projects the company has, the more meaningful the results.

2. Interview the project managers and compile a list of which problems have occurred.

3. Compare the list from each project and total up the problems.

4. Identify any trends and if the company is facing persistently common problems.

5. <u>REVIEW QUESTIONS WITH COMPANY</u>: This is important. The company may wish to alter the list of questions. Also, the company may be reluctant to have questions being asked about sensitive issues such a salary, promotions, career paths, etc.

6. <u>ESTABLISH INTERVIEW SCHEDULE</u>: The company will provide you with a contact person, usually from the personnel department. However, prior to contacting this person, the team should develop an interview availability schedule. <u>There must be two students per interview</u>. The class should plan for 20–30 interviews. The more interviews, the better the results. It is easier for the company to find people to be interviewed than for students who work during the day to reschedule their time.

    Try to request interviews at three levels of the organization: executive level, lower and middle management, and professionals. You may find that each level has a different opinion of the state-of-the-art of project management.

7. <u>CONDUCTING THE INTERVIEWS</u>: There should be two students present at each interview so that there will be agreement on what the interviewee actually said. With more than two students, the interviewee may feel threatened. It is important for the two students to review their notes immediately after the interview. Tape recording interviews is a good idea as long as the interviewee isn't threatened by it.

    Most interviews should be scheduled for 15–20 minutes. However, in reality, most interviewees may talk for an hour or more and provide the students with valuable information. The questions must be prioritized in case the students run out of time during the interview. A 20-minute interview will handle 5–20 questions.

    The interviewee must agree to have his/her name assigned to the comments. Any interviewees who request copies of the report should go through the company contact person.

8. <u>COLLATE THE DATA</u>: The complete interview process may take three to four weeks. By that time, the students may be far enough along in the course to deduce meaningful conclusions on their own.

9. <u>PRESENTATION TO INSTRUCTOR</u>: There are two reasons why presentation to the instructor is critical. The instructor must make sure, first, that the proper conclusions are identified, and, second, that the information in the report is valid.

The second item requires further comment. In one class, the instructor never reviewed the information in the report until being "called on the carpet" by a furious company officer. The students wrote in the report that one of the organization's senior project managers commented that the company has good "control" over the union workers. The union had seen the report and was very unhappy about this comment. The senior project manager swore that he had never made that comment, but both student interviewers stated otherwise. This problem should have been resolved prior to submittal of the report to the company.

10. <u>SUBMITTAL/PRESENTATION TO COMPANY</u>: Most of the time, a company is happy simply to receive a report because it is viewed as free consulting. However, there are situations where the company may want a formal presentation. The instructor must judge how much involvement he/she should have in the presentation. We must remember that this project is a reflection on the school as well as the students and the instructor.

THE STATEMENT OF WORK

Misinterpretation of the statement of work (SOW) can lead to expensive scope changes. Project managers with good communication skills can lower the barrier and perception screens that lead to SOW misinterpretation. Project managers who prepare statements of work are often just as guilty as those who must interpret a client's SOW in using improper language or making improper assumptions.

To illustrate the complexity of SOW preparation and interpretation, consider a customer that wishes to train approximately 200 employees in the principles of effective project management. This customer has prepared a statement of work, which is shown in Exhibit 1. Your company has been invited to bid competitively on this training program. The program is a firm fixed-price effort. All bids must be received by the customer no later than 30 days from today.

The SOW for this request for proposal appears in Exhibit 1. A "full" SOW would also include contractual terms and conditions, financial payment plans, and penalties. Analyze the "technical" SOW as it appears in Exhibit 1.

QUESTIONS

1. Are there any portions of the statement of work that can be misinterpreted or are simply vague? If so, what are they?

2. Is there any information missing from the statement of work that would be needed before effective pricing can be performed? If so, what information?

3. Can the government prevent this situation from recurring? If so, how?

4. Could part of the RFP review process be to evaluate bidders on previous performance on government contracts? Is this legal?

_____

_____

_____

_____

_____

_____

CRASHING THE EFFORT

Using the figure below and the table, answer questions 1, 2, and 3.

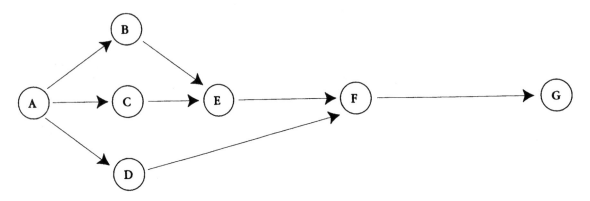

	Time (wks)		Cost ($)		
Activity	Normal	Crash	Normal	Crash	Crashing $ Per Week
AB	8	5	24,000	30,000	
AC	8	6	30,000	35,000	
AD	8	7	14,000	14,500	
BE	6	5	16,000	17,500	
CE	4	3	10,000	11,750	
DF	8	6	30,000	32,000	
EF	4	—	20,000	—	
FG	2	—	18,000	—	

1. Complete the column identified as "crashing cost per week."

2. What is the minimum additional cost needed if management wants the project completed in 19 weeks? In 18 weeks? In 17 weeks?

3. What is the minimum cost needed for a maximum crash and what is the length of the critical path?

THE AUTOMOBILE PROBLEM

You have recently been promoted to assistant project manager assigned to the proposal team that has been assembled to bid on an upcoming project. To get your feet wet, the project manager has given you the responsibility of determining the life-cycle cost of owning an automobile.

If your company wins the contract, you and your team will be required to make numerous trips to the client's site, which is located in a remote area. You estimate that the project, which is five years in duration, will require the purchase of an automobile. The customer has agreed to allow the bids to include this. Rough estimates indicated that you will travel 18,000 miles each year.

You must determine the total life-cycle costs of purchasing an automobile. After extensive research, you come up with the following data.

A. Purchase price is $19,000 with a salvage value of $2,500 at the end of the fifth year.

B. The automobile can be depreciated over three years using the straight line depreciation schedule.

C. Sales tax is 8 percent.

D. Insurance/year will be $870.

E. License/year will be $65.

F. Parking, garaging, and tolls will be $2,500 per year.

G. Gasoline is $1.40 per gallon, and the car is estimated to get 26 miles per gallon.

H. Major tuneups will cost $375 every 12,000 miles.

I. Minor tuneups will cost $140 very 6,000 miles.

J. Tires will cost $75 each, and all four tires must be replaced every 36,000 miles.

K. Alignment will be part of major tuneups.

L. Suspension system and brakes will be required every 24,000 miles at a cost of $220.

M. Mufflers will cost $130 (including installation) every 36,000 miles.

N. The electrical system (horn, washer, wipers, wiring, lighting, battery alternator) will require $650 every 44,000 miles.

QUESTIONS

1. Calculate the total cost of owning the automobile for each quarter of each year assuming that the car is sold at the end of the fifth year when the contract expires.

_____

_____

_____

_____

2. What assumptions must be made?

_____

_____

_____

_____

_____

3. On a sheet of graph paper, plot the cost/mile as a function of time.

4. What conclusions can be made?

_____

_____

_____

_____

5. If a loan were taken out to pay for the car, can the monthly interest expenses simply be included in the life-cycle costing? Explain you answer.

_____

_____

_____

_____

_____

6. Suppose you were asked to evaluate the option of leasing a car for five years rather than an outright purchase. What additional factors must be considered.

_____

_____

_____

LIFE-CYCLE COSTING

Warren Corporation has just completed a seven-year subcontracted project for a government agency. The project required the development of a high-technology piece of equipment during the first three years and operational support of the equipment in the field for the remaining four years. The customer was delighted that Warren Corporation was able to keep the operation and support cost at 39.9% of the total life-cycle cost rather than the originally estimated 60%, thus saving the government almost $19 million.

Table 1 shows the final Life-Cycle Cost Breakdown that was prepared by Warren Corporation at the closeout of the project. The project required the production of 25 units, all of which worked well. The problem facing Warren Corporation was the follow-on contract, which could conceivably reach a level of 6,000 units produced over the next 10 years.

Warren was able to use their existing production facilities for the first 25 units. The maximum production level that could be sustained in the existing plant would be 100 units per year. Warren would have to build a new manufacturing plant to support the production of 600 units per year.

The vice president for manufacturing has given you the task of preparing a detailed list of cost categories that must be looked at to assess the feasibility of building and operating a manufacturing plant for a 10-year lifetime. (Extension beyond 10 years is possible.)

Prepare a detailed work-breakdown structure of cost categories (as shown in Table 1) for the total life-cycle costing of the new plant. It is not necessary to use the same major headings as shown in Table 1.

TABLE 1

Cost Category	Constant Dollars	Percent Contribution (%)
Research & development cost	9,720,684	10.3
System product life-cycle management	1,415,634	1.5
Product planning	188,751	0.2
Product research	471,878	0.5
Engineering design	5,096,283	5.4
Design documentation	755,005	0.8
System/product software	660,629	0.7
System test & evaluation	1,132,504	1.2
Production & construction cost	42,752,154	45.3
Industrial engineering & operations	1,038,131	1.1
Analysis	34,163,974	36.2
Manufacturing	5,001,907	5.3
Construction	849,381	0.9
Quality control and initial logistics support	1,698,761	1.8
Operation & support	37,655,872	39.9
System/product operations	6,228,790	6.6
System/product distribution	4,813,157	5.1
Sustaining logistics support	26,613,925	28.2
Retirement & disposal cost	4,246,903	4.5
GRAND TOTAL	$94,375,613	100.0%

USING THE 50/50 RULE

The figure below identifies ten work packages that constitute a cost account. Answer the following questions using the figure below. (Budgets are represented in $1,000 increments.)

1. What is the total budget for this cost account?

2. Using the 50/50 rule, calculate BCWS and BCWP.

3. Assuming that ACWP is $45,000, calculate the cost and schedule variances.

4. Calculate CPI and SPI. Does the figure show a true representation of CPI and SPI? Why or why not?

5. Calculate EAC.

6. Are the "lengths" of the work packages correct? Explain your answer.

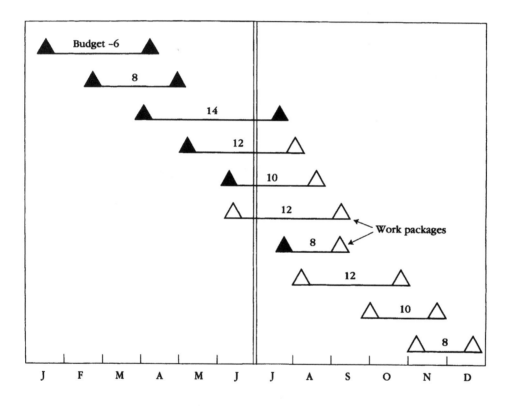

CONSTRUCTING PROCESS CHARTS

A small machine tool company prepares carbon rods for its clients. The production process is part manual and part mechanical. Each day, the company measures five of the rods to verify the stability of the process. The table below shows the measurements (in millimeters) for the last 10 working days.

DAY	READINGS (Millimeters)	RANGE (Millimeters)	$\overline{X}$ (Millimeters)
1	93 94 93 95 92		
2	89 93 94 92 91		
3	95 93 90 89 94		
4	90 93 94 92 92		
5	89 89 91 91 90		
6	92 93 93 95 95		
7	93 94 93 94 94		
8	89 93 93 92 91		
9	94 94 93 92 92		
10	95 94 93 94 92		
		$\overline{R}$ =	$\overline{\overline{X}}$ =

A. Complete the last two columns in the above table and calculate $\overline{R}$ and $\overline{\overline{X}}$.

B. Using the data/formulas in Appendix 1, calculate the upper and lower control limits for $\overline{X}$ and $\overline{R}$.

C. Plot the upper and lower control limits. $\overline{R}$, $\overline{\overline{X}}$, and the range and $\overline{X}$ data points on graph paper.

FORMULAS AND FACTORS FOR PROCESS CONTROL CHARTS

1. Factors for setting control chart limits (numbers are based upon three standard deviations assuming a normal distribution).

Sample Size	R Chart Factors		Factor to
n	Lower Limit, D_3	Upper Limit, D_4	Estimate, $o'd_2$
2	0	3.27	1.128
3	0	2.57	1.693
4	0	2.28	2.059
5	0	2.11	2.326
6	0	2.00	2.534
7	0.08	1.92	2.704
8	0.14	1.86	2.847
9	0.18	1.82	2.970
10	0.22	1.78	3.078

2. R Charts

    A. Upper Control Limit (UCL) = $D_4 \overline{R}$

    B. Lower Control Limit (LCL) = $D_3 \overline{R}$

3. $\overline{X}$ Charts

    A. Upper Control Limit = $\overline{\overline{X}} + \dfrac{3R}{d_2\sqrt{n}}$

    B. Lower Control Limit = $\overline{\overline{X}} - \dfrac{3R}{d_2\sqrt{n}}$

4. Terminology

    n = sample

    $\overline{X}$ = mean of each sample

    R = range of each sample (largest-smallest)

    $\overline{R}$ = average of all samples' ranges

    o' = estimate of the standard deviation = $\overline{R}/d_2$

CONSTRUCTING CAUSE-AND-EFFECT CHARTS AND PARETO CHARTS

Pacific Shuttle Airways has embarked upon a program to improve the quality of their services. More than 1,000 frequent fliers of Pacific Shuttle Airways were surveyed with the results shown in Exhibit 1.

1. Using the data in Exhibit 1, complete the Cause-and-Effect Diagram in Figure 1.

2. Using the data (i.e., groupings) from the Cause-and-Effect Diagram, prepare a Pareto Chart in Figure 2. Be sure to include only those elements that may be under the control of Pacific Shuttle Airways.

EXHIBIT 1
LATE FLIGHT DEPARTURE PROBLEMS

CAUSE	NUMBER OF RESPONSES
Late baggage to aircraft	340
Bad weather	240
Too few gate agents	66
Late cabin cleaners	21
Poor announcement of departures	12
Gate agents not motivated	103
Late aircraft arrival to gate	68
Late or unavailable cabin crews	21
Passengers bypass check-in counter	62
Late passenger cutoff too close to departure time	44
Late food services	37
Gate agents undertrained	116
Desire to "protect" late passengers	57
Late issuance of boarding passes	33
Late weight and balance sheet	40
Late or unavailable cockpit crews	19
Confused seat selection	66
Late fuel	35
Gate agents arrive late at gate	19
Late pushback tug	75
Checking oversized luggage	20
Desire to help company's income	37
Air traffic control delays	166
Poor gate location	130
Mechanical failures	26
Delayed check-in procedures	31
Gate occupied by a late departing plane	22

FIGURE 1
CAUSE-AND-EFFECT ANALYSIS

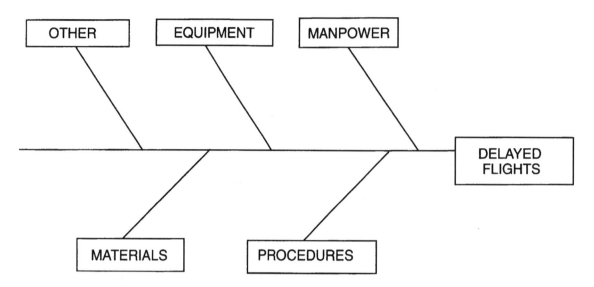

FIGURE 2
PARETO CHART

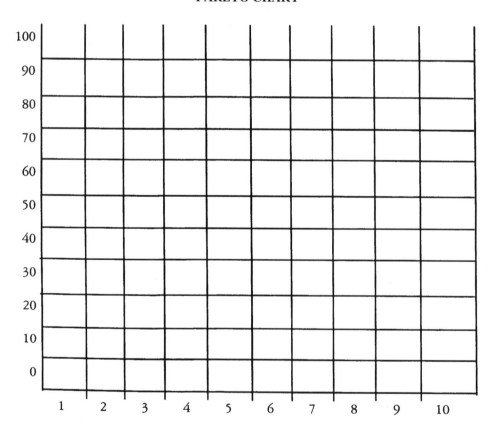

THE DIAGNOSIS OF PATTERNS OF PROCESS INSTABILITY, PART (A): $\overline{X}$ CHARTS

On the next page are five causes of instability in $\overline{X}$ and five figures. Match the five figures (1–5) with the five causes (A–E). Each figure is to be used only once.

CAUSE A: This indicates a gradual drift, probably caused by a machine part becoming worn down or possibly a gradual slippage in a machine setting.

        FIGURE: _____

CAUSE B: This could indicate different machine operators who perform differently. This could also indicate sticky machine settings.

        FIGURE: _____

CAUSE C: This indicates a cyclical process, perhaps from using different operators or different settings. Also, we should check the day of week or other cyclical variables.

        FIGURE: _____

CAUSE D: This could be a "freak" point in the process, perhaps due to faulty material or operator error.

        FIGURE: _____

CAUSE E: This could indicate a change in operators, machine settings, new materials, or an employee unfamiliar with the process specifications.

        FIGURE: _____

$\overline{X}$ CHARTS

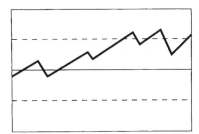

Figure 1

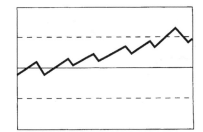

Figure 2

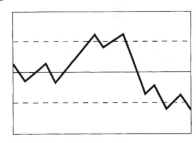

Figure 3

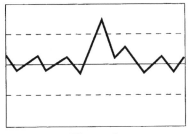

Figure 4

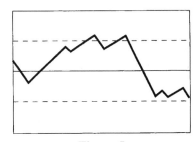

Figure 5

THE DIAGNOSIS OF PATTERNS OF PROCESS INSTABILITY, PART (B): $\overline{R}$ CHARTS

On the next page are five causes of instability in $\overline{R}$ and five figures. Match the five figures (1–5) with the five causes (A–E). Each figure is to be used only once.

CAUSE A: This indicates a machine that needs to be serviced, new lot of raw materials, a new raw material supplier, or an untrained employee.

　　　　FIGURE: _____

CAUSE B: This could be a new employee who is just learning the job.

　　　　FIGURE: _____

CAUSE C: This indicates a repetitive failure. If the problem occurs at the beginning of each week, it could be the result of the weekend shutdown.

　　　　FIGURE: _____

CAUSE D: This could be different employees with different skill levels and different training.

　　　　FIGURE: _____

CAUSE E: This indicates either a deterioration in the equipment signifying needed maintenance, or the operator was not adhering to performance requirements.

　　　　FIGURE: _____

$\overline{R}$ CHARTS

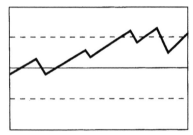

Figure 1

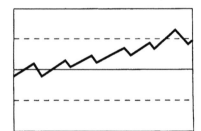

Figure 2

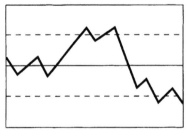

Figure 3

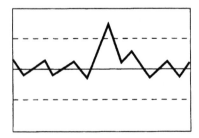

Figure 4

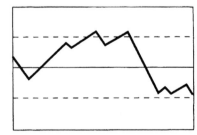

Figure 5

QUALITY CIRCLES

Below are four multiple choice questions on quality circles.

1. A quality circle is a group of people doing _____ work in the same area who _____ meet on a regular basis to identify, analyze, and solve problems in that area.

   a. Similar, voluntarily
   b. Similar or dissimilar, are required to
   c. Similar or dissimilar, voluntarily
   d. Similar, are required to
   e. Project, are required to

2. Quality circles work best if employees are initially trained in:

   a. Group dynamics
   b. Motivation principles
   c. Communications
   d. All of the above
   e. A and C only

3. Which problem-solving technique(s) has/have been proven to be effective for use in quality circles?

   a. Brainstorming
   b. Data collection and graphs
   c. Pareto Charts
   d. Cause-and-effect charts
   e. All of the above

4. Which of the following is false concerning the benefits of quality circles?

   a. Quality of products and services will improve
   b. Communications will improve
   c. Worker performance will improve
   d. Morale will improve
   e. All of the above are true

QUALITY PROBLEMS

1. Listed below are the eight stages in the industrial product development cycle. In which of these eight stages should effective quality management practices be adhered to?

   a. Marketing meets with the customer and determines the customer's expectations
   b. Engineering translates the customer's expectations into specifications
   c. Using the specifications, a bill-of-materials is prepared from which purchasing obtains the necessary parts and materials
   d. Manufacturing engineering selects the production process
   e. Manufacturing produces the item
   f. Inspection tests the item
   g. Shipping delivers the item to the customer
   h. Installation installs the item at the customer's location

2. Which of the following eight items are normal activities included in a TQM system?

    a. Preproduction quality evaluation
    b. Product and process quality planning
    c. Procurement quality planning
    d. Product and process evaluation and control
    e. Quality information feedback
    f. Quality training
    g. Management of quality control
    h. Special quality studies

3. Which of the following seven M's affect quality?

    a. Markets (expectations)
    b. Manpower (specialization)
    c. Money (profit margins)
    d. Management (of quality responsibility)
    e. Materials (high technology)
    f. Machinery (complexity/high technology)
    g. Methods (complexity)

4. Which of the following are true and which are false concerning the project manager's current perception of quality?

	TRUE	FALSE
A. Quality practices must extend throughout the entire project life cycle	_____	_____
B. Quality requirements should be established right at the onset (not too much or too little)	_____	_____
C. The project should be tailored to fit the organization's quality program	_____	_____
D. The quality function should be assigned to line managers rather than the project staff	_____	_____
E. The project quality manager must communicate, motivate, mentor, and help solve problems	_____	_____
F. Each team member should be involved through training, quality circles, suggestions, etc.	_____	_____
G. The project manager is responsible to ensure that the quality function is properly organized	_____	_____

Your Personal Learning Library

Write down your thoughts, ideas, and observations about the material in the chapter that may assist you with your learning experience. Create action items and additional study plans to assist you in enhancing your skills or for preparing to take the PMP® or CAPM® exam.

Insights, key learning points, personal recommendations for additional study, areas for review, application to your work environment, items for further discussion with associates.

_____

_____

_____

_____

_____

_____

_____

_____

_____

_____

_____

Personal Action Items:	
Action Item	**Target Date for Completion**

Chapter Twenty-Three

PROJECT MANAGEMENT CROSSWORD PUZZLES

Learning doesn't always have to be accomplished through tests and hours of study. Occasionally, some enjoyment should be added to the learning process to help you remember key points and to stay focused on your learning objectives. This chapter is intended to help you continue your project management learning experience through crossword puzzles that have been designed to not only enhance your knowledge of project management but also to provide you with an enjoyable yet somewhat challenging approach to the learning process. The crossword puzzles are a change of pace from the routine study process and will help you remember key terms and phrases through reasoning and determining relationships between topics and terms. These puzzles provide you with an opportunity to think in broader terms and across knowledge areas to better understand how all project management processes are integrated within a system. Use them for study on your own or in small study groups. You will find that as you complete each puzzle your knowledge will grow, you will find yourself reviewing your notes and the chapters within this workbook which will further enhance your total learning experience, and you will have a much greater understanding of how the many project management terms, principles, tools, and techniques are interrelated.

SCOPE MANAGEMENT

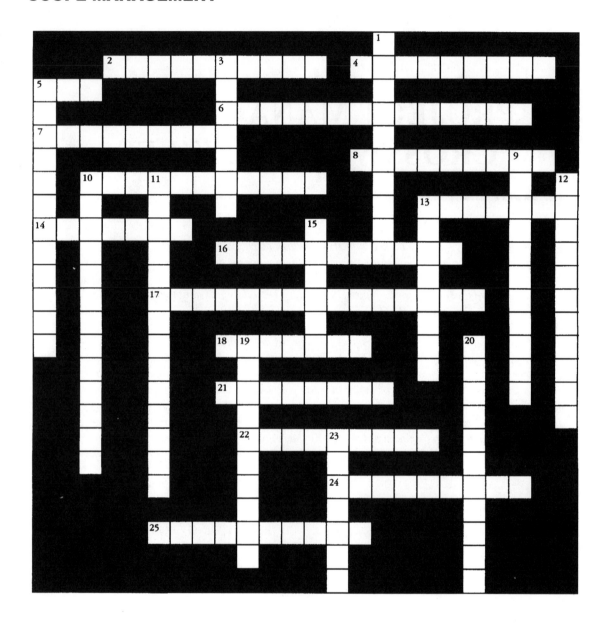

Across

2. CHOOSING BEST WAY TO ACHIEVE AN OBJECTIVE: _____ DEVELOPMENT

4. DEVIATIONS FROM A MEASURED PERFORMANCE PARAMETER

5. TASK-ORIENTED FAMILY TREE (ABBR.)

6. HEIRARCHICAL RELATIONSHIPS _____

7. PERIODIC DOCUMENTATION

8. LABOR, EQUIPMENT, FACILITIES, ETC.

10. RESTRICTIONS

13. SET OF STEPS TO MEASURE PROGRESS AND TAKE CORRECTIVE ACTION: _____ SYSTEM

14. REASON FOR PERIODIC REPORTING

16. RESEARCHING, ORGANIZING, AND COLLECTING DATA: _____ GATHERING

17. CONCEPTUAL DEVELOPMENT, SCOPE STATEMENT, WORK AUTHORIZATION, ETC.

18. EXECUTIVE, CUSTOMER, OR PROJECT _____

21. BINDING AGREEMENT FOR GOODS/SERVICES

22. TYPE OF REPORT THAT APPEARS ONLY WHEN OUT-OF-TOLERANCES OCCUR

24. SCOPE EXPRESSED IN TERMS OF OUTPUTS, RESOURCES, AND TIMING

25. "LOOSE" MANAGEMENT METHODOLOGIES

Down

1. TYPE OF PLAN DESCRIBING THE PROJECTS ORGANIZATION AND ADMINISTRATION

3. PROJECT, _____, OR PRODUCT LINE MANAGER

5. ELEMENTS ASSIGNED TO COST CENTERS FOR ACCOMPLISHMENT OF ACTIVITIES

9. A WAY TO INTEGRATE TIME AND COST REPORTING TO OBTAIN TRUE PROGRESS

10. CONTROLLING SCOPE CHANGE: _____ CONTROL

11. DOCUMENT DESCRIBING PROJECTS DELIVERABLES, TIMING, ORGANIZATION, AND PROCESS

12. EVALUATING DIFFERENT SOLUTIONS: _____ ANALYSIS

13. TERMINATION

15. DIRECTIONS OF A VARIATION

19. POLICIES _____, AND GUIDELINES

20. PROCUREMENT

23. GROUP OF RELATED ACTIVITIES THAT CONSUME RESOURCES AND HAVE CONSTRAINTS AND A DESIRED OUTCOME

TIME (SCHEDULE) MANAGEMENT

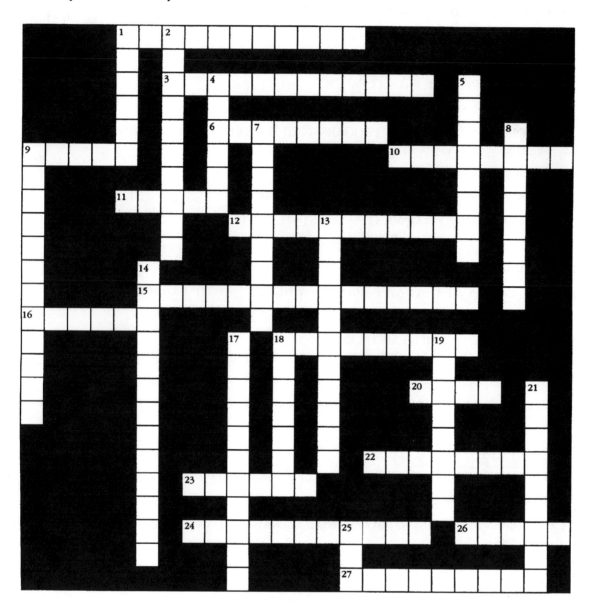

Across

1. CALCULATION OF ACHIEVEMENT
3. OVERLAPPING ACTIVITIES TO SHORTEN TIME
6. GRAPHICAL REPRESENTATION OF STATUS
9. ACTIVITY WITH ZERO TIME DURATION
10. PLANNED WORK VERSUS LIMITS ON RESOURCES
11. DEFINABLE SINGLE POINT IN TIME
12. AN ACTIVITY THAT MUST COME BEFORE ANOTHER ACTIVITY
15. RATIO OF ACTUAL TO PLANNED PERFORMANCE
16. TO REVISE BASED UPON CURRENT INFORMATION
18. EXAMINE WITH INTENT TO VERIFY
20. A CONTINUOUS, LINEAR SERIES OF ACTIVITIES
22. DIFFERENCE BETWEEN PLANNED AND ACTUAL
23. PLANNED ALLOCATION OF RESOURCES
24. ASSESSMENT OF CAPABILITY
26. INTERDEPENDENCY OF ACTIVITIES
27. SET OF RECOMMENDATIONS

Down

1. UNIFORM DIRECTIVES ISSUED BY MANAGEMENT
2. REWORK, REDEFINITION, OR MODIFICATION OF WORK
4. A FORM OF SUBPROJECT
5. TYPE OF PASS WHICH CALCULATES LATE START AND LATE FINISH TIMES
7. FACTOR WHICH LIMITS THE AVAILABILITY OF A RESOURCE
8. PREDICTION OF THE FUTURE
9. DISSEMINATION OF INFORMATION FOR APPROVAL OR DECISION MAKING
13. WRITTEN APPROVAL
14. PRECISE DESCRIPTION OF AN ITEM, PROCEDURE, OR RESULT
17. CALCULATION OF THE EARLY START AND FINISH TIMES
18. ACCEPT AS SATISFACTORY
19. OVERTIME WORK TO SHORTEN DURATION
21. A PREDETERMINED RESULT
25. A START-FINISH RELATIONSHIP

COST MANAGEMENT

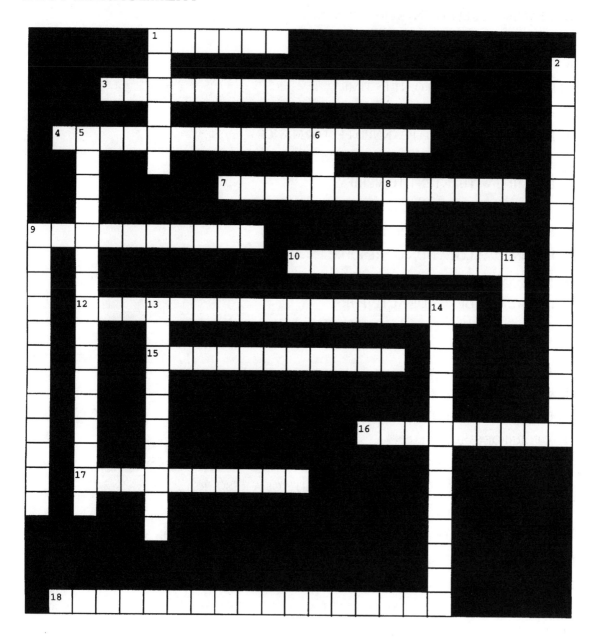

Across

1. PLANNED ALLOCATION OF FUNDS

3. NUMBERING SYSTEM ASSIGNED TO COST DATA

4. DEVELOPMENT OF CHANGES TO IMPROVE PERFORMANCE

7. COMPARISON OF INCOME TO MONIES SPENT

9. STATISTICAL OR _____ COST ESTIMATING

10. A STUDY DESIGNED TO PRESENT THE PROJECT TO OWNERS OR FUNDERS

12. PROJECT FUNDS RETAINED FOR MINOR CHANGES IN SCOPE

15. TYPE OF PLAN TO ALLOW FOR UNFORESEEN ACTIVITIES

16. SOURCES OF MONIES

17. (−5, +10%) ESTIMATE

18. ALLOWANCES FOR CHANGES

Down

1. PLANNED ALLOCATION OF FUNDS

2. ANALYSIS OF BIRTH-TO-DEATH PROJECT COSTS

5. (−25, +75%) ESTIMATE

6. BCWP/ACWP

8. EARNED VALUE

9. LABOR EFFICIENCY COMPARED TO A BASELINE

11. BCWP/BCWS

13. IDENTIFYING, MEASURING, AND RECORDING ACTUAL COST DATA

14. OPTIMIZATION OF COST TO PERFORMANCE

HUMAN RESOURCE MANAGEMENT

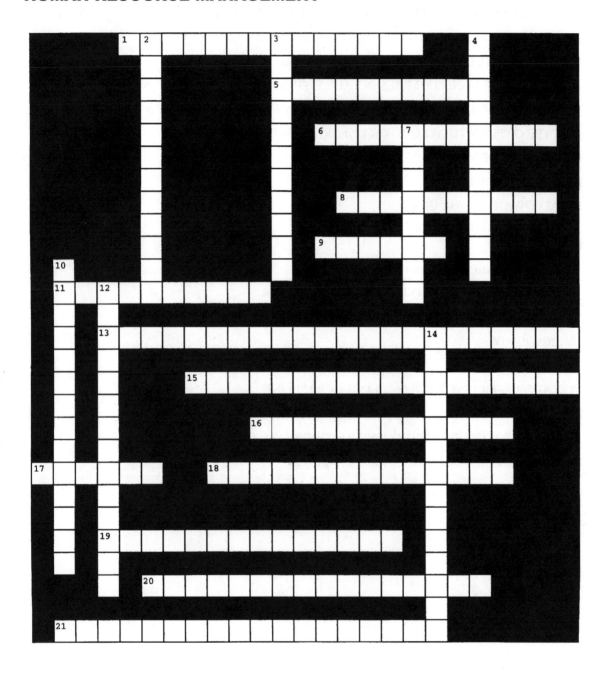

Across

1. INDIVIDUALS NOT DIRECTLY INVOLVED IN MAINSTREAM ACTIVITIES

5. ADVISING IN CAREER PLANNING, WORK REQUIREMENTS, OR QUALITY OF WORK

6. FORMALIZED SYSTEM FOR DEALING WITH GRIEVANCES

8. INDUCING ONE TO WORK TOWARD AN OBJECTIVE

9. MULTIDIMENSIONAL ORGANIZATIONAL STRUCTURE

11. DISTRIBUTION OF AUTHORITY OR RESPONSIBILITY

13. MANAGEMENT OF NONFINANCIAL COMMITMENTS TO EMPLOYEES

15. PROCESS TO DEAL WITH DISAGREEMENTS

16. MATCHING EMPLOYEE TO TASKS BASED UPON SKILL LEVEL

17. WAGE AND _____

18. WRITTEN OUTLINE OF ACTIVITIES OF A JOB

19. MATRIX, EXPEDITOR, OR PURE PROJECT _____ STRUCTURES

20. PROJECTING RESOURCES NEEDED OVER TIME

21. PROCESS TO DEAL WITH DISAGREEMENTS

Down

2. INFLUENCING A GROUP TO WORK TOGETHER

3. ACKNOWLEDGMENT

4. BARGAINING

7. DEVELOPMENT OF SPECIFIC SKILLS

10. WAGE AND SALARY _____

12. NEGOTIATING AND BARGAINING WITH THE LABOR FORCE

14. INFLUENCING A TEAM TOWARD GOAL ACCOMPLISHMENT

COMMUNICATIONS MANAGEMENT

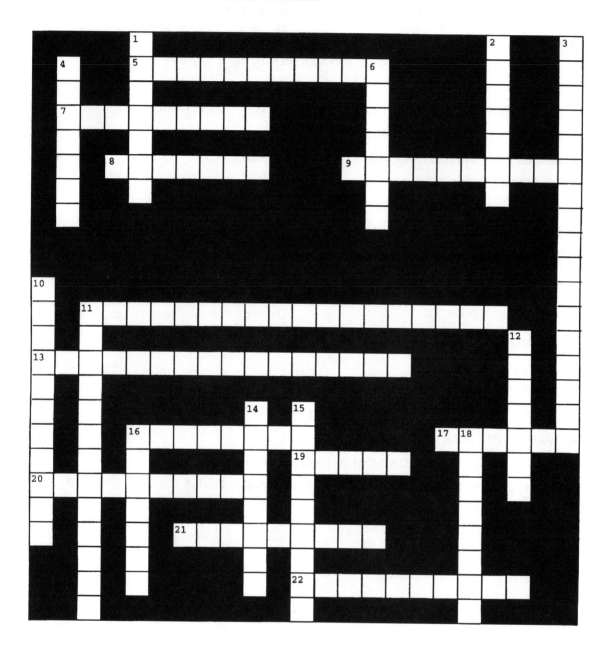

Across

5. SURROUNDINGS
7. PAYING ATTENTION TO WHAT IS SAID
8. LITERARY FORM
9. RETREATING FROM A CONFLICT
11. SEEKING SOLUTION TO A PROBLEM OR DISAGREEMENT
13. AUTHORITARIAN, DISRUPTIVE, JUDICIAL, ETC.
16. MOVE BY ARGUMENT
17. SET OF MORAL PRINCIPLES OR VALUES
19. VISUAL, AUDITORY, TACTILE, ETC.
20. INFORMATION EXCHANGE AMONG GROUPS
21. RELATED TO THE SENSE OF SMELL
22. GIVE AND TAKE AGREEMENT

Down

1. TYPE OF NONVERBAL COMMUNICATION
2. WRITTEN, GRAPHICAL, OR PICTORIAL TRANSMISSION OF INFORMATION
3. TYPE OF NONVERBAL COMMUNICATIONS
4. PATTERN OF HUMAN KNOWLEDGE
6. RELATING TO THE SENSE OF TOUCH
10. SOUND BUT PRUDENT JUDGMENT
11. FACE-TO-FACE MEETING TO OBTAIN A SOLUTION
12. A WIN-LOSE CONFLICT RESOLUTION MODE
14. A SYSTEMATIC MEANS OF COMMUNICATION
15. LANGUAGE USED TO ACHIEVE A DESIRED EFFECT
16. SERIES OF ACTIONS LEADING TO AN END
18. SEND OR CONVEY

RISK MANAGEMENT

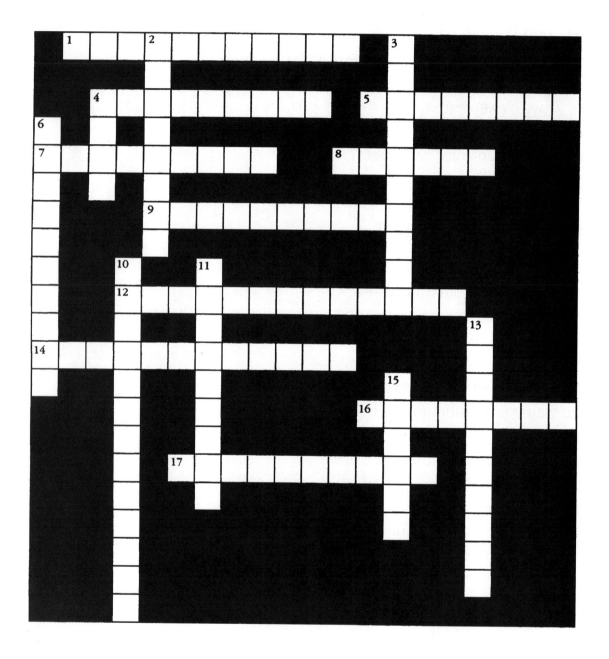

Across

1. CERTAINTY, RISK, AND _____

4. RISK ASSUMPTIONS

5. DETERMINING PROBABILITIES AND THE CONSEQUENCES: RISK _____

7. PURE, AS IN RISKS

8. RISK: UNCERTAINTY, DAMAGE _____ (PLURAL)

9. EXAMINING A RISKY SITUATION: RISK _____

12. SUMMATION OF PAYOFFS (2 WORDS)

14. FAVORABLE OUTCOMES

16. METHODS TO DEVELOP OR CONTROL RISKS: RISK _____

17. PHASE WITH GREATEST OVERALL RISKS

Down

2. RISKS OUTSIDE OF PROJECT MANAGER'S CONTROL

3. PHASE WITH LEAST AMOUNT OF COST AT STAKE

4. TOTAL ADVERSE EFFECTS

6. REVISION OF CONSTRAINTS TO REDUCE THE UNCERTAINTY WITHOUT ALTERING THE OBJECTIVES

10. RISK IDENTIFICATION: WBS _____

11. RISK PREVENTION OR CONTROL

13. TRANSFERRING RISK TO ANOTHER PARTY

15. SOURCE OF RISK DANGER

QUALITY MANAGEMENT

Across

2. _____ COSTS: TRAINING, VENDOR SURVEYS, PROCESS STUDIES, ETC.

8. QUALITY IS DEFINED BY THE _____

9. 85 PERCENT OF QUALITY ISSUES DUE TO _____

11. TYPE OF NONCONFORMANCE COST (2 WORDS)

13. QUALITY _____ IS STATEMENT OF WHAT, NOT HOW

15. PATTERN OF DATA POINTS

16. TYPE OF NONCONFORMANCE COST (2 WORDS)

17. A BALDRIGE CRITERION

Down

1. _____ COSTS: EVALUATION OF PROCESS

3. COMMON AND SPECIAL CAUSE

4. EXPERIMENTAL DESIGNER

5. QUALITY PIONEER

6. PART OF USL

7. PROCESS ABNORMALITY IS _____ DATA POINTS

10. QUALITY PIONEER

12. QUALITY PIONEER

14. PART OF LCL

PROCUREMENT MANAGEMENT

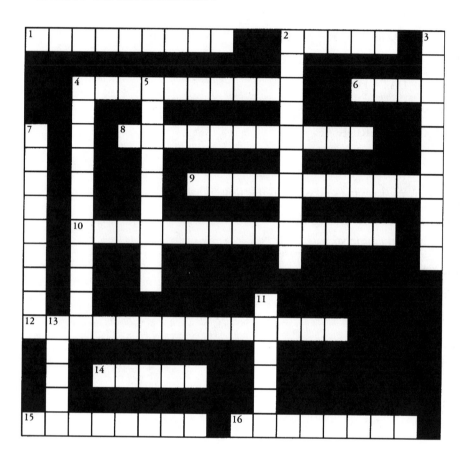

Across

1. TIME AND _____

2. FIRM _____ PRICE CONTRACT

4. _____ TO BID

6. _____ SOURCE PROCUREMENT

8. PROCUREMENT

9. PERIOD OF _____

10. CONTRACT _____ CYCLE

12. NOT MEETING THE SPECIFICATION

14. CONTRACT _____ CYCLE

15. REQUEST FOR _____

16. TYPE OF PAYMENT

Down

2. TYPE OF SPECIFICATION

3. ORDER OF _____

4. REQUEST FOR _____

5. TYPE OF CONTRACT

7. REQUEST FOR _____

11. AUTHORIZED DEVIATION FROM SPECS

13. WORK STOPPAGE _____

SUMMARY

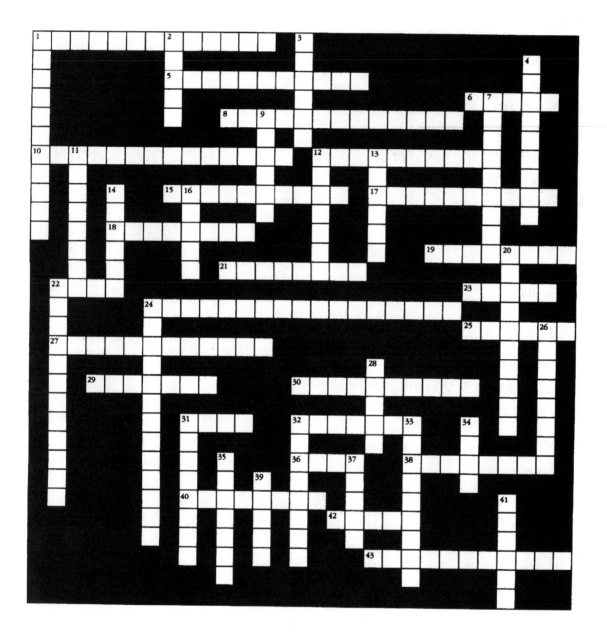

Across

1. PROCESS OR EXCHANGING INFORMATION

5. _____ BIDDING

6. CONFIGURATION CONTROL _____

8. SUGGESTIONS WITHOUT EVALUATION

10. AUTHORITY PLUS RESPONSIBILITY

12. ADM, PERT, AND _____ NETWORKS

15. _____ FACTORS (I.E., PREDICTIONS)

17. PERT ARROWS

18. _____ & DEVELOPMENT

19. CONFIGURATION IDENTIFICATION DOCUMENTS

21. PART OF BCWS

22. A SET OF ACTIVITIES DIRECTED AT PRODUCING A DESIRED END

23. NOTIFICATION TO BIDDER OF ACCEPTANCE OF BID

24. ONE OF MASLOW'S NEEDS

25. PERT NOMENCLATURE (PLURAL)

27. HARDWARE, SOFTWARE, OR REPORTS PROVIDED TO THE CUSTOMER

29. ONE WHO INSPECTS RECORDS FOR HONESTY AND ACCURACY

30. _____ OR SIMULTANEOUS ENGINEERING

31. PART OF ACWP

32. A FEE PAID BY THE INSURED TO THE INSURER FOR SOME TYPE OF PROTECTION

36. _____ LEAD ITEMS

38. EXPERIENCE OR _____ CURVES

40. AN AGREEMENT ENFORCED BY LAW

42. EARLY OR LATE _____

43. WHAT-IF OR _____ PLANNING: ACCOUNTING FOR FUTURE EVENTS THAT ARE NOT ANTICIPATED

Down

1. TIME, COST, AND PERFORMANCE _____

2. PROJECT LIFE _____

3. LETTER-OF- _____ : SUCH AS OBLIGATION FOR LONG-LEAD ITEMS

4. PART OF BCWP

7. THE END RESULT THAT BUSINESS SETS FOR ITSELF

9. PART OF ACWP

11. _____ PATH

12. COMBINATION OF HUMAN AND NONHUMAN RESOURCES PULLED TOGETHER TO ACHIEVE AN OBJECTIVE

13. BAR OR MILESTONE _____

14. OVERHEAD PLUS NONREIMBURSABLE COSTS

(*continues*)

Down (*continued*)

16. PERT NOMENCLATURE: _____ TIME
20. THE ART OF INFLUENCING OTHERS
22. ACTUAL RATE OF OUTPUT PER UNIT TIME OF WORK
24. TECHNICAL REQUIREMENT
26. ON-THE-JOB _____
28. LOW BIDDING A JOB WITH THE EXPECTATION OF FUTURE WORK ADDITIONS
31. PENALTY OR _____ POWER
32. RIGID GUIDELINES
33. A SIGNIFICANT EVENT IN A NETWORK
34. _____ PACKAGES
35. COST VERSUS _____ ANALYSIS
37. BAR OR _____ CHARTS
39. PURCHASE OR WORK _____ NUMBER
41. LIQUID, CURRENT, OR FIXED _____

Sixteen Points
(Based on Dr. Kerzner's 16 Points to Project Management Maturity)

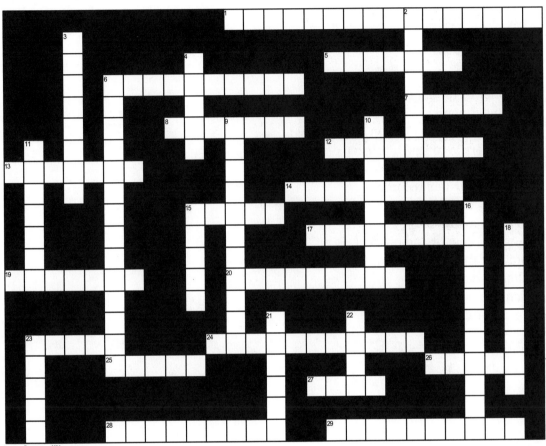

www.CrosswordWeaver.com

ACROSS

1 Time, cost, and scope
5 strengthen, provide aid,
6 a belief or system of beliefs
7 use jointly or in common
8 achievement of goals
12 measured accomplishments
13 lessons _____
14 people and materials
15 confidence or faith in someone
17 capable, achieves results, functions as expected
19 gathering of people for a purpose
20 work or act together
23 Sum of the work

24 A way of doing something
25 fast, speedy
26 Description of an objective
27 _____ manager (has direct responsibility for functional resources
28 To acknowledge
29 Regularly occurring

DOWN

2   supplier of funding
3   computer application
4   concentration of attention and energy
6   Right person to lead a project
9   exchange information
10 maintain written records
11 quantify a result or evaluate performance
15 to teach a new skill
16 tangible item or end product
18 State of full development
21 alter or modify
22 Scheme or intention
23 an acquired ability

SOLUTIONS TO CROSSWORD PUZZLES

ANSWER KEY
SCOPE MANAGEMENT

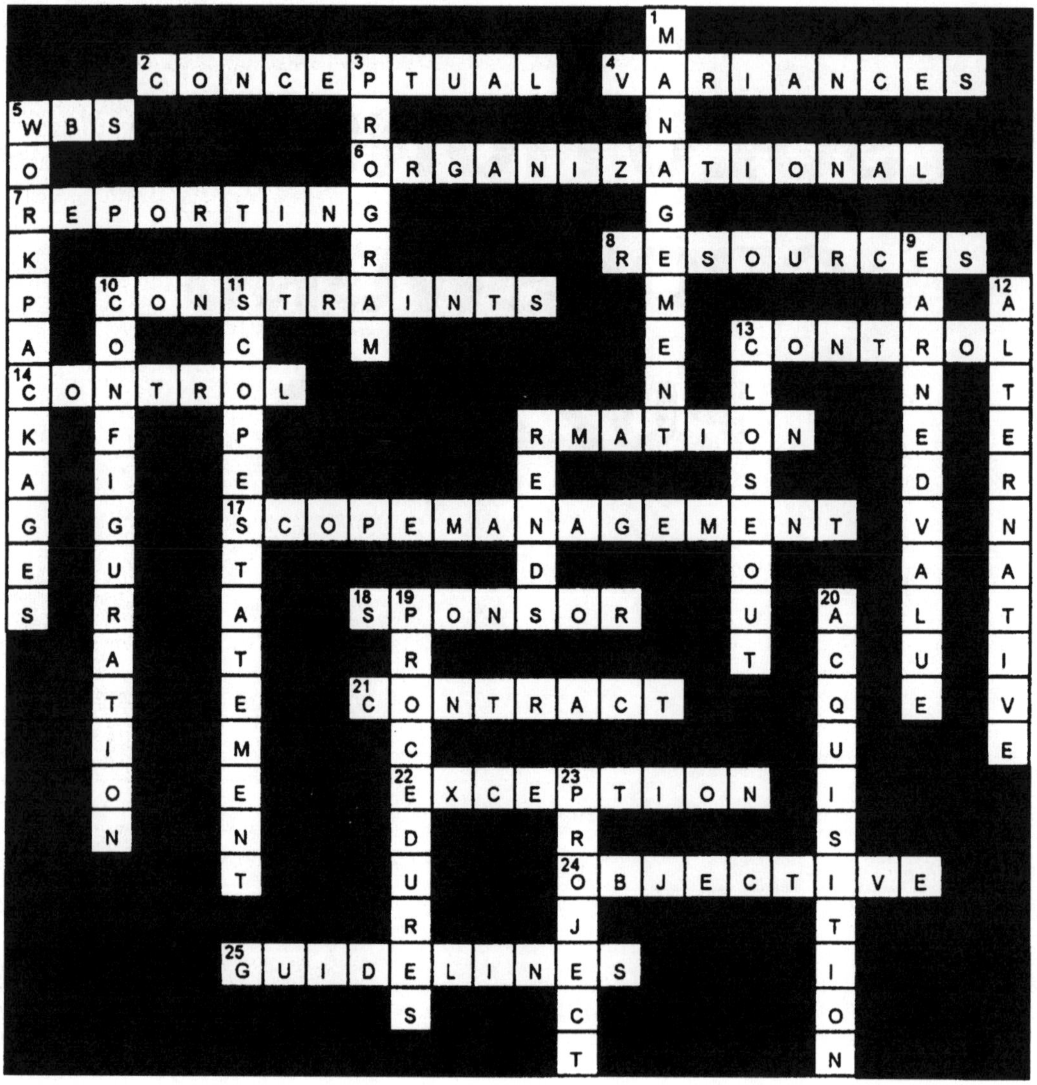

A completed crossword puzzle grid with the following answers:

Across/Down answers:
- CONCEPTUAL
- VARIANCES
- WBS
- ORGANIZATIONAL
- REPORTING
- RESOURCES
- CONSTRAINTS
- CONTROL
- CONTROL
- RMATION
- SCOPEMANAGEMENT
- SPONSOR
- CONTRACT
- EXCEPTION
- OBJECTIVE
- GUIDELINES

ANSWER KEY
TIME (SCHEDULE) MANAGEMENT

ANSWER KEY
COST MANAGEMENT

ANSWER KEY
HUMAN RESOURCE MANAGEMENT

Across:
1. STAFF PERSONNEL
5. COUNSELING
6. ARBITRATION
8. MOTIVATING
9. MATRIX
11. DELEGATION
13. BENEFITS ADMINISTRATION
14. (TRATION)
15. CONFLICT MANAGEMENT
16. JOB PLACEMENT
17. SALARY
18. JOB DESCRIPTION
19. ORGANIZATIONAL
20. MANPOWER PLANNING
21. CONFLICT RESOLUTION

Down:
2. TEAM BUILDING
3. RECOGNITION
4. NEGOTIATION
7. TRAINING
10. ADMINISTRATION
12. BORRELL
14. TEAM MOTIVATION
18. JOBS

ANSWER KEY
COMMUNICATIONS MANAGEMENT

Across

5. ENVIRONMENT
7. LISTENING
8. WRITTEN
9. WITHDRAWAL
11. CONFLICT RESOLUTION
13. MANAGEMENT STYLES
16. PERSUADE
17. ETHICS
19. MEDIA
20. NETWORKING
21. OLFACTORY
22. COMPROMISE

Down

1. G
2. DISPLAY
3. FACIAL EXPRESSION
4. CULTURE
6. TACTILE
10. COMMONSENSE
11. CONFRONTATION
12. FORCING
14. LANGUAGE
15. STATUS
18. TRANSMIT

ANSWER KEY
RISK MANAGEMENT

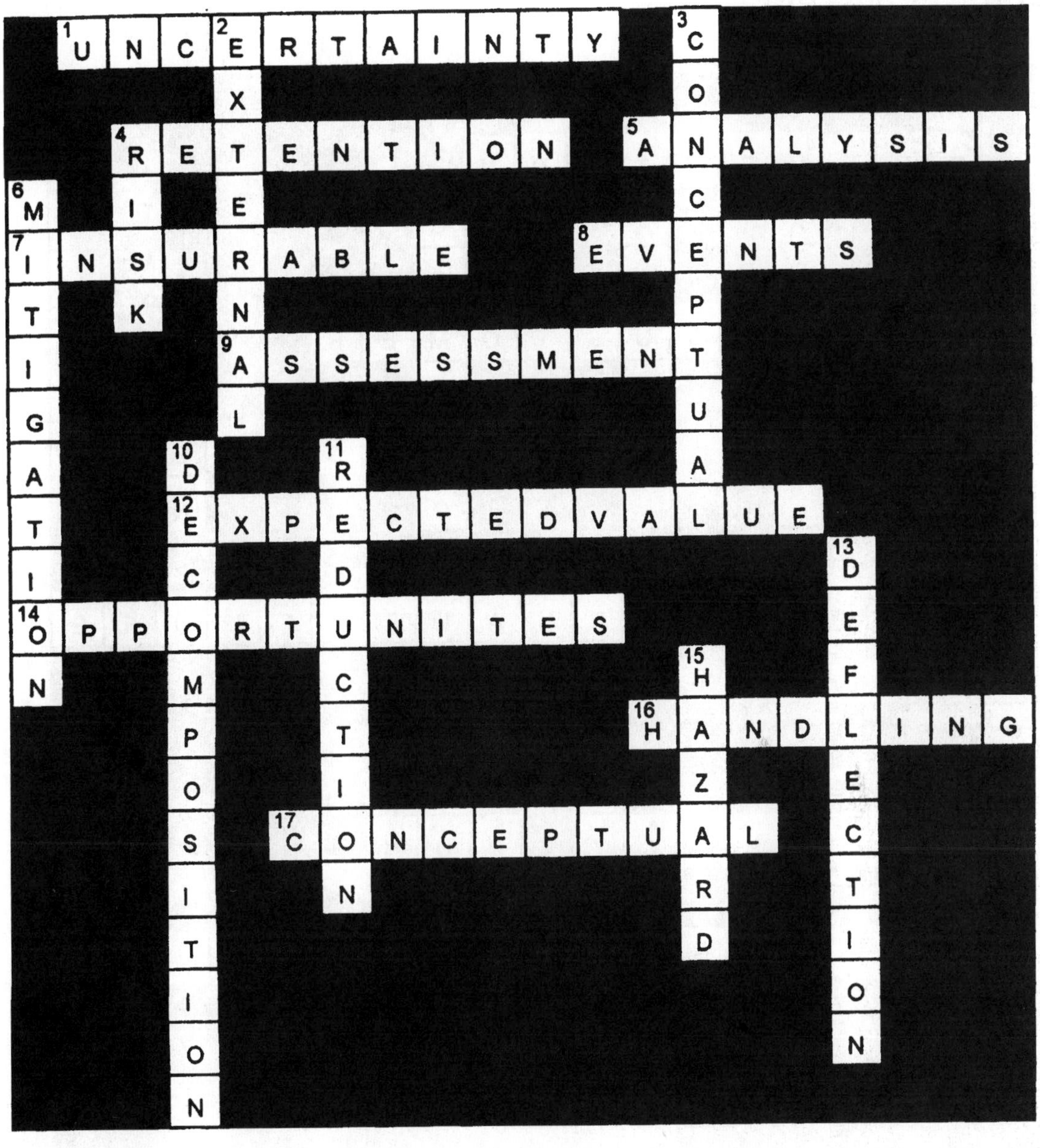
1 UNCERTAINTY
2 EXTERNAL
3 CONCEPTUAL
4 RETENTION
4 RISK
5 ANALYSIS
6 MITIGATION
7 INSURABLE
8 EVENTS
9 ASSESSMENT
10 DECOMPOSITION
11 REDUCTION
12 EXPECTEDVALUE
13 DEFLECTION
14 OPPORTUNITES
15 HAZARD
16 HANDLING
17 CONCEPTUAL

ANSWER KEY
QUALITY MANAGEMENT

Across
2. PREVENTION
8. CUSTOMER
9. MANAGEMENT
11. INTERNAL FAILURE
13. POLICY
15. RUN
16. EXTERNAL FAILURE
17. LEADERSHIP

Down
1. APPRAISAL
3. VARIABILITY
4. TAGUCHI
5. DEMING
6. LIMIT
7. SEVEN
10. CROSBY
12. JURAN
14. CONTROL

ANSWER KEY
PROCUREMENT MANAGEMENT

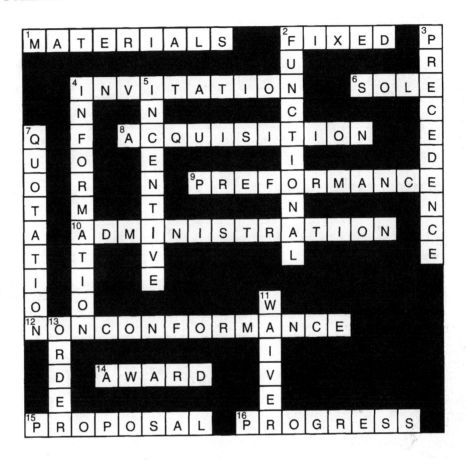

ANSWER KEY
SUMMARY

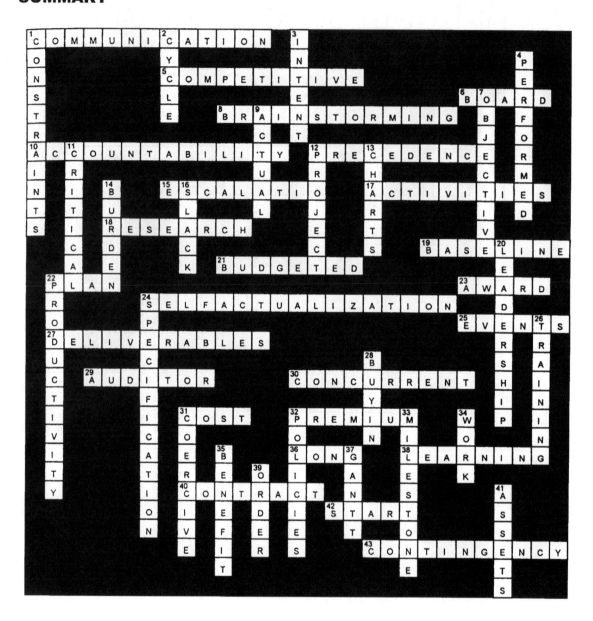

Answer Key
Sixteen Points

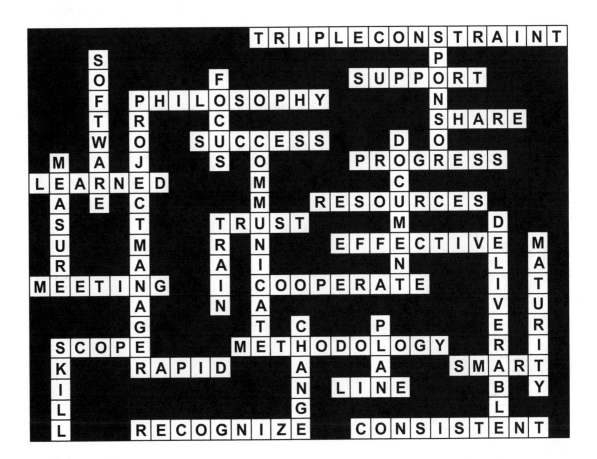

Your Personal Learning Library

Write down your thoughts, ideas, and observations about the material in the chapter that may assist you with your learning experience. Create action items and additional study plans to assist you in enhancing your skills or for preparing to take the PMP® or CAPM® exam.

Insights, key learning points, personal recommendations for additional study, areas for review, application to your work environment, items for further discussion with associates.

_____

_____

_____

_____

_____

_____

_____

_____

_____

_____

_____

Personal Action Items:	
Action Item	**Target Date for Completion**

Chapter Twenty-Four

PMP® AND CAPM® EXAM REVIEW

The Project Management Professional or PMP® exam is based on the processes within the domains of project management known as: Initiating, Planning, Executing, Monitoring and Controlling, Closing, and Professional and Social Responsibility. It is a four-hour, 200 multiple-choice question, computer-based exam (in the U.S.) that is intended to test a candidate's knowledge of project management and the ability of the candidate to apply that knowledge in many different situations and scenarios. Candidates for the exam must apply to the Project Management Institute® and complete a detailed and rigorous application process. For information about the exam and how to apply as a candidate visit the PMI® website at www.pmi.org.

Eligibility for the Exam

Applicants with a high school diploma must have 7,500 hours in a position of responsibility leading and directing specific tasks, 60 months of project management experience, and 35 hours of formal project management education.

Applicants who hold a baccalaureate degree (or equivalent) are required to have 4,500 hours leading and directing specific tasks and 36 months of project management experience.

A complete PMP® Credential Handbook is available from PMI®.

The PMP® exam is a comprehensive exam that challenges the knowledge of the candidate through situational- and scenario-based questions that are designed to simulate actual project management experiences. The exam is based in part on the *PMBOK® Guide, (generally the current edition),* but also includes questions that go beyond the framework and processes described in the *PMBOK® Guide.* Knowledge of leadership techniques, organizational structure, managerial theory and practice, team motivation, conflict resolution, negotiating, problem solving, scheduling techniques including forward and backward pass, contract sharing ratios, ,and the application of earned value analysis will be tested along with the processes explained in the *PMBOK® Guide.*

The Certified Associate in Project Management (CAPM®) certification examination is a knowledge-based exam that uses *A Guide to the Project Management Body of Knowledge (PMBOK® Guide) current edition,* as the source of questions. The CAPM® exam is a three-hour, 150 multiple-choice question examination. The CAPM® exam focuses on the fundamental knowledge of project management, and all questions are derived from the *PMBOK® Guide.* PMI® offers a basic knowledge assessment on line which will assist in determining your level of knowledge of project management and provide you with a basis for further study and preparation.

Eligibility for the CAPM® Exam

Option 1—High school diploma or equivalent and 1,500 hours of work on a project team
Option 2—High school diploma and 23 contact hours of project management education

Applicants must complete the CAPM® Credential Application online or via a downloadable form available from the PMI® website.

Preparing for the Exams

The PMP® exam has been developed for experienced project managers and project team members who are in decision-making and leadership positions and those who influence project decisions. The CAPM® exam has been developed for project managers who plan to continue their careers as project managers and have not met the criteria for education, experience, and credentials to apply for the PMP® exam. The CAPM® is an excellent step toward the PMP certification.

Although the PMP® and CAPM® exams are designed for different levels of qualification and experience, the following information will assist in the study and preparation for each exam.

Working the Workbook

This section of the workbook is intended to provide a summary review of the key points of project management, the processes within the *PMBOK® Guide*—Fourth edition, and highlight the elements of each knowledge area that may be tested on the exam.

☑ **Study Note: It is important to obtain a copy of the *PMBOK® Guide* and read it thoroughly. Many people will tell you that you should read it at least three times to absorb the information provided. The *PMBOK® Guide* contains an enormous amount of information about project management and is referred to as the standard for developing a project management methodology. The key to successful study is to understand how each knowledge area is interrelated and that all knowledge areas of the *PMBOK® Guide* are integrated in the project planning process and the management of a project.**

A Basic Study Plan

- Read every chapter of the *PMBOK® Guide* including the first three chapters that describe the project management context and framework.
- Read chapters 1 through 3 in one study session. These chapters are essential in developing an understanding of how the *PMBOK® Guide* is arranged.
- Read each of the remaining chapters in succession (the recommendation is to read one chapter at each study session). The chapters contain a large amount of information and you can easily become overwhelmed and lose focus. Patience and a steady pace is your best strategy. Make notes as you review each chapter and review them often. Use additional references to obtain further details about the information provided in the *PMBOK® Guide*. The *PMBOK® Guide* does not provide detailed and comprehensive explanations of many important project management terms, tools and techniques,
- Ensure that you understand the flow charts that appear in each chapter and the relationships and flows of the inputs, tools and techniques, and outputs of each process within a knowledge area.

- Review the overview charts often and learn to recognize the inputs, tools and techniques, and outputs. Concentrate on why and input is important, how a tool or technique is applied and how the resulting output becomes an input to another process. It is not essential to memorize the *PMBOK® Guide*. The study goal is not memorization, but understanding rationale and application of the knowledge, tools and techniques
- As you progress through the *PMBOK® Guide* you will begin to see a repetition of many processes. The processes are interrelated and the outputs of one process will frequently become the inputs to another process. This repetition emphasizes the integrative nature of project management.
- The chapter entitled "Project Integration Management" refers to many processes that you may not be familiar with. They are described in later chapters. Make a note of unfamiliar terms and look for them in later chapters.
- When you have completed all chapters, review the glossary and then reread "Project Integration Management." This will help to further embed the knowledge obtained from the other chapters.
- Test yourself frequently with practice questions and try to relate the processes explained in the *PMBOK® Guide* with actual projects you have worked on or are currently working on. Not every process will be applicable to your projects but it will help with your study if you can relate the processes to projects whenever possible.

Preparing for the Exam

- Obtain the PMP® exam certification package from www.pmi.org.
Review and verify that you are qualified to take the exam.
- Obtain a copy of the current *PMBOK® Guide* (there is usually an overlap period when a new edition of the *PMBOK® Guide* is released. Check with PMI to determine which edition is appropriate for study.
- Read each chapter of the *PMBOK® Guide*. Take notes to further increase retention but do not try to memorize each section. Learn to apply the knowledge through practical application.
- When you find a term or subject that you are not familiar with, research the topic to ensure you have a full understanding.
- Read the *PMBOK® Guide* one chapter at a time. There is a lot of information to absorb.
- Obtain additional books for study: Recommendation: *Project Management: A Systems Approach to Planning, Scheduling, and Controlling*, 10th Edition, Harold Kerzner, Ph.D., John Wiley & Sons, 2006.
- Plan to study at least one hour each day and plan for a minimum of 40 to 60 hours.
- Plan to take a formal class in project management. You need 35 hours of formal training.
- Take note of all formulas that are associated with project planning especially in project selection (NPV), time management—PERT weighted average formula, Earned Value Management, Standard Deviation, Communications channels, CPM-Forward and Backward pass, expected monetary value, cost –benefit ratio, contract sharing ratios. Most of the formulas are not available or explained in the *PMBOK® Guide*. Use other resources to obtain this information.
- Obtain practice exams.
- Start a study group and develop your own questions and presentations.
- Get plenty of rest the night before the exam.
- Last-minute cramming is not recommended!

Questions to Consider Regarding the Various Topics Discussed in This Workbook

This workbook has introduced you to several project management processes, tools and techniques. To further your knowledge, ask yourself the following questions:

1. Select a process, tool or technique, or a specific concept and ask yourself "How can I apply this concept, process, or technique to my current environment?" Example: Creating a WBS for a new project. How

would you obtain buy-in from team members? How would you start the process of developing the WBS? What preparatory work would you be required to perform?

2. If you cannot apply the concept, ask yourself why not? What is preventing the use of the concept? Where is there resistance? Why is there resistance? Why is the concept not applicable to your environment?

3. What changes must be introduced within the organization to gain acceptance of the process or technique?

These questions will assist you in developing a more complete understanding of the processes, tools and techniques described in the material and will prepare you to become an advocate for project management.

Knowledge Review—Key Topics, Learning Points, and Exam Tips

Reasons why an organization needs project management.

- To determine roles and responsibilities
- To identify and establish priorities
- To manage deadlines and schedules
- To determine and assign accountability and authority
- To ensure proper documentation—determining what project documentation is needed and to avoid over-documentation
- To provide a common methodology that will improve efficiencies in resource management and estimating
- To provide effective and timely status and communications
- To effectively control changes to the project plan and to the objectives of the project
- To manage multiple projects more effectively
- To reduce rework and eliminate redundant work
- To establish consistency in the planning and executing of project work
- To manage risk effectively
- To balance the competing demands of scope, quality, schedule, budget,, resources and risk
- To capture and share lessons learned for continuous improvement
- To develop best practices and focus on continuous improvement
- To ensure targeted value of the project is achieved

PMBOK® Guide—Key Learning Points of Chapters 1, 2, and 3:

Enterprise environmental factors Internal and external factors that influence a project's success. Examples: organizational culture, industry standards, infrastructure, human resources.

Project A temporary endeavor that is unique, constrained by resources, planned, executed, and controlled. A project produces one or more deliverables. Project work is similar to operations work but the difference is that project work is temporary and operations work is considered ongoing.

Progressive Elaboration Proceeding in steps, increasing in detail as the project moves forward through phases.

Project Management The application of knowledge, skills, and tools to meet stated requirements. This includes setting objectives, balancing competing demands of time, cost, scope, and quality.

Relationship of the 9 Knowledge Areas Three key words -Integration, interrelationships, and inter-dependencies (all 9 knowledge areas are part of an integrated system). Consider project management to be a system of components, each significantly influencing the other parts of the system.

Project Environment It is important to be aware of cultural differences, the international and political environment, and the physical environment that a project team may be working with. These items are directly associated with professional and social responsibility.

Interpersonal Skills The successful project manager can effectively manage project teams and interpersonal relationships through communication, influencing skills, leadership, motivation, negotiating skills, conflict management, and problem solving.

Program Generally considered to be comprised of several projects that are related in some way and are managed in a coordinated way to obtain benefits that would not be possible if they were managed separately. Programs do not have a definitive end date and tend to be ongoing for longer, sometimes undetermined, time frames.

Portfolio Management A collection of projects and programs and other work that are grouped together to facilitate effective management of that work to meet strategic objectives

Project Management Office An organizational body or entity assigned various responsibilities related to the centralized and coordinated management of those projects under its domain.

9 Knowledge Areas of the *PMBOK® Guide* Integration, Scope, Time, Cost, Risk, Human Resources, Communications, Quality, and Procurement Management. All 9 knowledge areas are integrated and have an inter-relationship. The knowledge areas are part of a project management system The inputs for one process are, in many cases, the results or outputs from a process associated with another knowledge area. Review the overview charts found in each chapter of the *PMBOK® Guide* to observe the relationships between knowledge areas.

Relationships to Other Knowledge Areas Project management has many similarities to General Management. General management skills include financial management and accounting, purchasing and procurement, contracts and commercial law, strategic planning, tactical planning, operational planning, organizational structures, safety practices, and information technology. Project management requires knowledge of these areas to ensure effective planning and execution. Project Management may also include Application Area Knowledge such as software development, construction, aerospace and automobile design where there may be a requirement to possess knowledge of the technology and terminology associated with the specific discipline to successfully manage the project.

Deliverable A tangible, verifiable work output. Deliverables can be identified within each project phase and reviewed for acceptance prior to the start of the next phase.

Phases and Project Life Cycles Projects are divided into phases to make the project more manageable. Phases also allow project managers to manage progress, make decisions regarding project continuation (go–no-go decisions) and provide an opportunity to review project progress and performance and to assess project risks more effectively. There are three basic phase to phase relationships – sequential, overlapping, and iterative.

Projects have greater risk and a low probability of success at the start. In the early stages of the project the scope of the project has not been fully elaborated, and the details required to develop an effective plan have not been completely defined. Uncertainty is extremely high resulting in a perception of greater project risk. As the project is progressively elaborated the risk associated with the project is reduced as more information becomes available resulting in more reliable plans.

Stakeholders People or organizations directly involved in or in some way impacted either positively or negatively as a result of the project. Stakeholders can be considered positive (support the project) or negative (impede the project). Project managers should consider the risks and potential impact of the negative stakeholder (whether internal or external to the organization). Key stakeholders include: Project Manager, Sponsor, Customer, Project team, the PMO—Project Management Office, the performing organization, operations management, sellers and business partners, the end user, and contractors..

Organizational Structures There are three basic organizational structures: Functional, Matrix , and Projectized. The matrix structure can be further defined as one of three possible sub types – Weak matrix where the project manager has little to no authority,, balanced matrix where the project manager and function manager have equal levels of authority, and strong matrix where the project manager has a greater level of authority than the functional manager. Consider the advantages and disadvantages of each type of structure. Example—The project manager authority in each type of organizational structure is different and may have an impact on how the project manager influences the project. In the functional structure the project manager has very little authority and acts as a coordinator or expeditor. In the strong matrix the project manager has greater influence and has the authority to direct functional managers.

Organizational process assets Processes and procedures for conducting work. Examples: standardized guidelines, templates, project closure guidelines, change control procedures

Key Management Skills The key skills associated with managing and with project management include: leadership,, communications, organizing, team building, conflict resolution, planning, administration..

Three dimensions of Project Management Competency Knowledge – what is known about project management, performance – what the project manager can accomplish by applying project management knowledge. Personal –behavior, core personality characteristics, attitude, and the ability to manage and lead a team.

Socio-Economic Influences and the Project Environment This is an important item to consider when planning and managing a project. A project manager must be aware of the environmental influences that may impact project planning and execution. These include government regulations, cultural differences, customs, international conditions, language, political sensitivities, regional holidays, physical geography, and time zones. These items are also associated with the professional and social responsibility domain of project management.

Plan–Do–Check–Act The Shewhart Chart adopted by Deming is an underlying concept of project management and relates to the five major project management process groups – Initiating, planning, executing, monitoring and controlling and closing described within the *PMBOK® Guide*.

Project Management Process Groups Initiating, Planning, Executing, Monitoring and Controlling, and Closing. It is important to note that these are not phases. These processes are present in each phase of the project life cycle.

Rolling Wave Planning A form of progressive elaboration where near term work is planned in detail at a low level of the WBS and work scheduled for the future is planned at a high level. As work is completed in one phase or period, the work of the next phase or near term is planned in more detail.

The initiating process group includes the development of the project charter and the identification of project stakeholders. The project charter is the document that defines the initial set of requirements, authorizes the use of organizational resources and formally assigns the project manager to the project. Identification of stakeholders is necessary to ensure that the appropriate requirements are established and to gain an understanding of the level of influence of the stakeholders involved.

PROJECT INTEGRATION MANAGEMENT

Each knowledge area in the *PMBOK® Guide* consists of several processes. These processes are managed through inputs, tools and techniques, and identified outputs. The processes within each knowledge area can be illustrated by using a flow chart to demonstrate how the key processes are interrelated and produce the key outputs. Each knowledge area is part of the project management system and all knowledge areas are integrated within the system.

The main process groups within Project Integration management are: Develop Project Charter, Develop Project Management Plan, Direct and Manage Project Execution, Monitor and Control Project Work, Perform Integrated Change Control, Close Project or Phase.

Study Tip: Review the overview chart provided in each knowledge area chapter of the *PMBOK® Guide* and follow through from each process group input to the tools and techniques and then to outputs and observe where the outputs become inputs to the next process. You will see some outputs appear in the process groups of other knowledge areas. You will also notice that many inputs and tools and techniques, are used in multiple processes within each knowledge area. Take time to review why an input is important and how tools and techniques will be applied to produce an output.

The Project Charter A document that formally recognizes the project and authorizes the project manager to apply organizational resources. The charter establishes the initial set of requirements and identifies the project stakeholders. To develop the project charter it is important to refer to the business need driving the project, the product to be produced, and the strategic objectives of the organization. .If a contract is involved, review the terms and conditions agreed upon.

Project Statement of Work Narrative description of products and services to be supplied by the project.

Study note: Always consider the enterprise environmental factors and organizational process assets that may affect the project planning process.

Project Selection Methods Prior to the development of the project charter the organization will consider which projects would be most beneficial for the organization. Benefit measurement methods, a benefit cost analysis (should we do it?), and feasibility study (can we do it?) are generally conducted before a project is approved. Many organizations consider the tangible (profit, cost savings) and intangible (good will, morale, customer satisfaction) benefits when selecting projects. Selection methods may include calculation of NPV—Net Present Value (time value of money). An important formula to remember is: Present value $PV = FV/(1 + r)^n$ .where PV is present value, FV is Future value, r = rate, n is the number of years. Other forms of project selection methods include Internal Rate of Return (IIR)— an iterative process that will determine the rate or cost of capital at which the project NPV would equal zero, breakeven analysis, pay-back period, ROI—Return on Investment, and cost- benefit ratio.

The PMIS (Project Management Information System) includes all of the systems you will use to obtain project data, analyze data, and distribute information to the project stakeholders. The Project Management Information System is any system or group of systems working together to gather, store, and distribute information about your project. Examples—Time Reporting System, Accounting System, and Project Software.

The 3rd edition of the *PMBOK® Guide* included a Preliminary Scope Statement as an output associated with Integration Management. The preliminary scope statement is the first step in defining the project in detail and provides an initial view of project objectives and what must be accomplished. It includes the project and product objectives, project boundaries, constraints, assumptions, initial risks, schedule milestones, requirements, and deliverables. The preliminary scope would be reviewed by stakeholders and eventually updated and approved. In some organizations the scope statement goes through several stages of vetting before final approval.

The Project Management Plan This is basically the guide that will be used by the project manager and team and includes the actions required to integrate the many subsidiary plans such as cost management, change control, quality management, and scope management into the total or integrated project plan. The project management plan is expected to change as the project progresses due to the uncertainties and changes in requirements and other factors.

Direct and Manage Project Execution   Performing the work described in the project management plan such as creating deliverables, obtaining project staff, generating project data and reports, procuring resources. The outputs of the direct and manage project execution include project deliverables, work performance information, change requests, corrective actions, preventive actions, defect repairs, document updates

Monitor and Control Project Work This includes tracking, reviewing, and managing project progress. Comparing actual performance to planned performance. Reporting project status.

Integrated Change Control Monitoring results, taking corrective action, managing change requests, review and approval of configuration changes. A change control process is used to minimize scope changes and reduce or eliminate scope creep. Approved changes require plan updates (documentation updates), implementation of the change, validation that the change has been successfully implemented, and recording of lessons learned. The 9 knowledge areas are managed in an integrated manner with an understanding that a change in one area can impact any or all of the other knowledge areas. Consider the impact of the change before implementing the change. Changes such as a schedule slip or a cost overrun may be experienced. To determine that a change of this nature has occurred compare the baseline with actual results. When a change is approved and must be implemented it is important to determine when to make a change and how to introduce the change to minimize the impact on ongoing operations, the current flow of project work or other projects.

Closing the Project or Phase Project plans should include a process for closing out a project or a phase of a project. There are two major procedures for project closure: Administrative closure—project reviews, verifying project financials, analyzing project success or failure, lessons learned, and Contract closure—the actions required to close contractual agreements and ensure that terms and conditions have been met. The closing process should include a plan that will satisfy project completion requirements and define exit criteria.

Historical Records These records provide an opportunity to learn from past projects. Use lessons learned and experience to reduce project related problems and recurring project issues.

Organizational Policies Consider the policies and procedures defined by your organization and how they can impact the planning of your projects. Example—Hiring process, budget policies, selection process, etc.

Constraints and Assumptions Identify the limitations or boundaries you must operate within. Determine what assumptions have been made or can be made about the project, Assumptions in this context are items we can believe to be true, real, or certain for planning purposes. All assumptions should be validated to avoid serious risk situations.

Organizational Strategies Consider the goals and objectives of your organization and how your project impacts or supports those goals. Make sure you can link your project to the organizational goals.

Project Plan The approved document that provides the baseline for executing and managing the project.

Work Authorization System The process that ensures the right work is done at the right time in the right sequence.

Change Control Board A team or group designated or empowered to review and determine the value of a change and to approve or deny change requests.

Configuration Management A process that will ensure that configuration changes (changes to features, functions, physical dimensions) are managed appropriately and approved to prevent or reduce the risks of additional cost, scope changes, or other impacts to the project.

PROJECT SCOPE MANAGEMENT

Ensuring that the project includes all work and only the work required to complete the objectives of the project.

The project manager and project team will become involved in the Scope Planning Process Creating a project scope management plan that describes and documents how the project scope will be defined, verified, controlled, and how the WBS will be developed. The scope management plan is a subsidiary plan to the project management plan.

There are two types of scope to be aware of:

Product Scope – the features, functions, and physical characteristics of the product service, or result of the project.

Project Scope The work that must be completed to produce the product , service, or result

Scope management includes the following major processes:

Collect Requirements Defining and documenting the needs of the stakeholders. Remember stakeholders can be affected negatively or positively by the results of the project.

Key terms Interviews, focus groups, facilitated workshops, group creativity techniques, ,group decision making techniques, observations, surveys

Define Scope Developing a detailed description of the project scope and the product or products to be delivered.. Proper scope definition will enable the project team to develop effective plans to achieve the desired results.

Key terms Expert judgment, product analysis, alternatives identification, facilitated workshops. Other terms associated with scope definition: Stakeholder Analysis, Project Scope Statement, Project Deliverables, Product Acceptance Criteria.

Project Objectives should be "SMART" Specific, Measurable, Attainable, Realistic, and Time-based

Create WBS The project team jointly develops the work breakdown structure using the process of decomposition. . The WBS is a hierarchal grouping of project components, tasks and activities. The WBS is defined to a level of detail that will enable the project team to plan and manage the project effectively. The lowest level deliverable of the WBS is the work package. The work package includes the activities that must be completed to produce higher level deliverables..

Key terms Decomposition – the process of breaking the WBS into smaller parts, WBS Templates – Templates may be created from previous projects and can accelerate the planning process., WBS Dictionary – provides detailed information about elements of the WBS, Scope Baseline – the defined project work that must be completed.

Other Breakdown Structures OBS (Organizational Breakdown Structure), BOM (Bill of Materials), and RBS (Risk Breakdown Structure)

Verify Scope The process of determining the correctness of the work completed through inspections and reviews of project work results. This means making sure that what was expected to be produced was actually produced. Scope verification leads to formal ACCEPTANCE. This is an important item to remember—connect scope verification with formal acceptance of the project deliverables

Key terms Inspection, Acceptance, Requested Changes

Control Scope Monitoring the project to ensure changes are managed through an approval process. Use of change requests and a configuration control process. Approved changes require updates to the plan and documentation of lessons learned.

Key terms Change Control System, Variance Analysis

PROJECT TIME MANAGEMENT

The processes required to ensure timely completion of the project. Project time management is centered around the activities defined in the WBS. In addition, time management is associated with effective use of resources, managing personal time, delegation of work, anticipating risks, and balancing the needs of the stakeholders.

Define Activities Working with the project team to identify the specific schedule activities that must be performed. This process is an extension of the development of the WBS. Work packages include the project activities that must be completed to produce the project deliverables. Inputs include: Enterprise environmental factors, organizational process assets, scope baseline,. Other inputs may include: historical data, WBS, constraints, assumptions, expert judgment.

Activity Definition Tools and Techniques Decomposition: breaks the WBS down further into work packages which include activity lists. Rolling wave planning, templates, expert judgment

Key terms Rolling Wave Planning, Activity List

Sequence Activities Remember the three dependency types: Mandatory (hard logic), Discretionary (soft logic), and External dependency (factors outside the project that may affect the project). This process produces a network diagram that will display predecessor/successor relationships or logical relationships.

Key terms Precedence diagramming method (PDM) or activity on the node, arrow diagramming method or ADM (this method is not commonly used. The PDM is the prevalent form of network diagramming for projects), Types of Dependencies—Mandatory, Discretionary, and External. , Leads (acceleration of time between activities and Lags – delays in time between activities.

Estimate Activity Resources Determining the number and types of resources needed for the project.

Tools and techniques – expert judgment, alternatives analysis, published estimating data, bottom-up estimating, project management software.

Key terms Activity Attributes, Bottom-up Estimating (uses the WBS to provide details at the work package level for more reliable estimates), Resource Breakdown Structure

Estimate Activity Duration Remember to determine the availability of the resources needed for the project, the number of resources needed, and assess resource capabilities. Consider historical information, analogous estimates, reserve time, and contingencies.

Tools and techniques – Expert judgment, analogous estimating (comparing a project to a similar project), parametric estimates (using mathematical algorithms) three point estimating (using optimistic, pessimistic and most likely values), reserve analysis (determining if the appropriate reserves or contingencies have been included)

Key terms Activity attributes, resource calendars, Analogous estimating, , Parametric estimating, Three Point Estimates, Reserve Analysis

Develop Schedule To properly develop a schedule that can be managed effectively, you must consider availability of resources, capability and skills of the resources, number of resources available, the resource calendar (including international calendar issues), vacation time, sick time, external issues (weather), and lead or lag requirements

Key terms CPM (Critical Path Method) – uses four types of logical relationships: finish to start, start to start, finish to finish, start to finish. , Schedule Compression – fast tracking and crashing, Network Diagram, Resource Leveling, Leads,, Lags, Schedule Baseline

Schedule Development Tools and Techniques Network Analysis, Schedule Compression which includes: *Fast Tracking,* which means overlapping tasks. This technique generally adds risk to the project schedule so it is important to assess the risks before fast tracking and *Crashing* of the critical path, the reducing of activity durations by adding cost generally through the addition of resources.

Key terms Resource Leveling, Critical Chain Method, What-if Scenarios, Bar Charts, Milestone Charts, Network Diagrams

Control Schedule This includes determining the status of the project schedule, influencing the factors that create schedule changes, determining if a schedule change has occurred, and managing actual changes. This involves the review of project reports, performance, change requests, and the use of a schedule change control system. Control includes the analysis of variances, taking corrective action, and documenting lessons learned.

Control Schedule Tools and Techniques Performance reviews, variance analysis, resource leveling, what-if scenarios, adjusting leads and lags, schedule compression.

Key terms Variance analysis, resource leveling, progress reporting, Schedule Change Control System

Important Time Management Items to Remember:

PDM—Precedence Diagram Method. There are 4 logical relationships: FS, FF, SS, SF
AOA—Activity on the arrow: may use dummy activity. Uses only FS relationship. The dummy activity may be a factor in determining the critical path
CPM— Critical Path Method. Determines the early start and early finish and late start and late finish of each project activity
Slack—Also known as *float*. A measure of flexibility in a network path. There is generally no float or slack on the critical path.
Free Float – the amount of time a task may slip without affecting succeeding activities
Total Float – the amount of time an activity may slip without affecting the project end date
Float ownership – the path owns the float, not the individual activity
Forward pass—Determines early start and finish dates
Backward pass—Determines late start and finish dates

PERT Weighted Average formula—(Optimistic + 4 Most Likely + Pessimistic)/6

Standard deviation = (b – a)/6 (Pessimistic – Optimistic divided by 6)

Path convergence – multiple activities that precede an activity. In the forward pass process the early start of the succeeding is activity is determined by the activity with the greatest value early finish.

PROJECT COST MANAGEMENT

Project Cost Management includes the processes that are used to determine project costs, establish the project budget, allocate costs to each phase of the project life cycle, monitor project expenses, and to control project costs through analysis of variances to the approved project budget.

Typical project costs include: Human resources (salaries), Materials and Equipment, Capital Costs, rent, leases, travel expenses, overhead expenses, recurring costs (electricity, water), training, and insurance.

The three major processes associated with Project Cost Management are:

Estimate Costs Developing an approximation of the project costs. Use of Analogous, Parametric (Top-down estimates), and Bottom-up Estimating (also known as engineering or grass roots estimates). Make sure you know the differences and consider the advantages and disadvantages of each type.) *Example:* Low accuracy of top-down estimate. The cost-estimating process produces cost estimates and a cost management plan.

Inputs to cost estimating   Scope baseline, project schedule, human resource plan, risk register, enterprise environmental factors, organizational process assets. Consider why each of these inputs are necessary to develop reliable estimates. Example: the risk register provides a list of potential risk events that may impact project cost and may require cost contingency reserves to be added to the budget.

Key terms Analogous estimating (use of similar projects to develop a top down estimate), Parametric Estimating (use of mathematical algorithms),, and Bottom-up Estimating (Use of the WBS), Cost of Quality – costs associated with conformance to quality requirements and non-conformance, Reserve Analysis – analyzing the planned reserves, , level of accuracy – rounding data to a prescribed precision level,, units of measure – hours, weeks, lump sum,, control threshold – allowable variance from a plan, rules of performance measurement – use of earned value management. Vendor bid analysis – review of bids to determine accuracy, fairness, consistency within the discipline.

Determine Budget Aggregating the estimated costs of all individual project activities to establish the project cost baseline, Creating the time-phased budget used to monitor project cost performance.

> **Inputs to Determine Budget** Basis of the estimates (details about how estimates were determined, scope baseline, project schedule, resource calendars, contracts
>
> **Key terms** Cost Aggregation- total of all project costs, Reserve Analysis, Cost Baseline, Time-Phased Budget- allocation of project costs to each phase of the project.

Control Costs Monitoring for changes or variances in project costs and taking corrective action. This also includes the management of change requests, using performance data, and producing measurements for analysis. Using earned value management to identify variances and trends.

> **Inputs to Control Costs** Project management plan, funding requirements,, work performance information, organizational process assets.
>
> **Key terms** Variance Management, Cost Change Control System, Performance Measurement Analysis, Earned Value Management

Other Important Cost Management terms

Learning Curves As units are produced, the time and cost to produce the unit decreases. Productivity improves through repetition. The rate of improvement decreases during the process.

Life-Cycle Costing The total cost of a product from research and development to salvage or retirement.

Future Value Formula $FV = PV (1 + r)^n$ FV= Future value   PV = Present Value   r=rate or cost of capital   n= years

Present Value Formula $PV = FV/(1 + r)^n$

Net Present Value The sum of the present values minus the initial investment. If NPV is greater than or equal to zero the project is generally considered acceptable. The greater the NPV the more acceptable the project will be financially.

IRR Internal Rate of Return. This is an iterative process to determine the interest rate or cost of capital at which the NPV becomes zero. The IRR generally must meet or exceed a predetermined "hurdle rate" established as part of project selection criteria. This process will provide an organization with a relaibel assessment of the return on investment of a project. IRR is compared with other projects during the selection process.

Payback Period Period of time required to recover the initial investment. It generally does not consider the time value of money.

Breakeven The point at which cash outflow and inflow are equal.

Sunk Cost Money already spent and not considered in decisions to go forward.

Indirect Costs Costs that are not directly associated with the project but that can impact the overall budget. Examples – Overhead, building maintenance, benefits, organization operational costs

Direct Costs Costs directly and clearly associated with the project. Labor and material are considered direct project costs.

BCWP Budgeted Cost of Work Performed This is the estimated or planned cost of the work that has been completed at the time of measurement

BCWS Budgeted Cost of Work Scheduled This the planned cost of the work that should have been completed at the time of measurement.

ACWP Actual Cost of Work Performed This is the amount that was paid for the work that was performed at the time of measurement.

Earned Value: Special note The PMP exam may use the acronyms EV (Earned Value) instead of BCWP, PV (Planned Value) instead of BCWS, and AC (Actual Cost) instead ACWP.

Earned Value Formulas:

EV (BCWP) – AC (ACWP) = Cost variance (CV) (a negative cost variance = budget overrun)
EV – PV (BCWS) = Schedule variance (SV) (a negative SV means the project or work package is behind schedule)
EV/AC (ACWP) = Cost performance index (CPI) (a measure of efficiency in managing the project budget)
EV/PV (BCWS) = Schedule performance index (SPI) (a measure of efficiency in managing the project schedule)

A performance index (PI) that equals 1 means there is no variance and the project or work package is is performing as planned. A PI greater than 1 indicates superior performance. A PI less than 1 indicates a a schedule slippage or a cost overrun.

EAC Estimate at Complete $EAC = AC^C + ETC$ (Estimate to Complete) – this formula is used when new estimates are developed for remaining work.

$EAC = AC^C + BAC - EV^C$ where BAC is the Budget at Complete   This formula is used when variances are seen as atypical and similar variances are not anticipated to occur again.

$EAC = AC^C + (BAC - EV^C / CPI^C$ Used when current variances are seen as typical and will be seen in the future

PROJECT QUALITY MANAGEMENT

Project Quality Management includes the processes that determine quality policies, quality objectives and responsibility of quality during project planning and implementation.

Project quality management emphasizes Customer Satisfaction, Prevention over Inspection, Continuous Improvement, and management responsibility.

Plan Quality Determining what quality standards are relevant to the project.

Inputs to Plan Quality Scope baseline, stakeholder register, cost performance baseline, schedule baseline, risk register, organizational process assets, enterprise environmental factors.

Key terms Cost benefit analysis, benchmarking, control charts, statistical sampling, proprietary quality management methodologies, design of experiments, Cost of Quality, Quality Metrics, Process Improvement Plan, flowcharting

Perform Quality Assurance The application of the processes and systematic quality activities that will ensure project success and the achievement of project requirements. The auditing of quality requirements and results to ensure appropriate levels of quality are being achieved are or meeting standards

Key terms Quality audits, process analysis

Perform Quality Control Monitoring and recording the project results to determine compliance with established and agreed upon requirements and assess performance. Quality control also includes recommendations for changes to resolve quality related issues.

Key terms Cause and effect diagram, control charts, flowcharting, Pareto Chart (80/20 rule), run chart, histogram, scatter diagram, statistical sampling, inspection, defect repair review

Other Quality Related Terms and Key Points

Quality Gurus W. Edwards Deming – Plan Do Check Act, 85% of quality issues require management involvement, Statistical analysis. Joseph Juran – Fitness for use, Legal issues associated with quality, Philip Crosby – Quality is free, cost of quality is associated with non conformance, zero defects is possible. Taguchi – design of experiments, quality should be engineered into a product,

Key Components of Quality Management Customer satisfaction, prevention of over-inspection, management responsibility, continuous improvement

The Customer Defines Quality The project team takes implied needs and turns them into stated needs to meet the customer's expectations. The project manager has ultimate responsibility for quality on a project.

Quality Policy Developed by the organization. It is an indication of the intentions of an organization as it relates to quality. The organizational quality policy is usually established by upper management.

Prevention Over Inspection It is generally less costly to prevent errors than to inspect and identify errors that must be corrected.

TQM Total Quality Management: An organizational approach to quality that starts at the top management level and includes all levels of employees. The focus is on continuous improvement through training and awareness about quality.

Quality Planning Inputs include the scope statement, product description, the company quality policy. Identifies which quality standards are relevant to the project.

Benefit Cost Analysis Determining the appropriate level of quality to meet an acceptable return on the investment while meeting customer needs.

Benchmarking Comparing desired or best in class performance to existing performance. Identifying gaps and developing corrective actions..

Flowcharting Diagramming a process to identify all steps, identify where gaps exist, and to identify redundant work. Flow charts are used to gain an understanding of the overall operation or flow of work.

Design of Experiments Introduced by Taguchi, the general principle is that quality should be built into (designed into) a product, not inspected into it.

Cost of Quality There are two major components of the Cost of Quality - Cost of conformance: prevention and appraisal and Cost of nonconformance—internal failure (rework) and external failure (customer dissatisfaction).

Quality Assurance Systematic activities implemented to ensure that the project will satisfy quality standards.

Quality Control Monitoring project results to determine if they comply with quality standards.

Prevention Keeping errors out of the process and from the customer.

Sampling Two types of sampling - Attribute and variable Attribute is a pass or fail approach, with no flexibility in accepting results. Variable allows for some tolerance and variance is measured on a scale..

Inspection Walk through, testing, reviews.

Control Charts Rule of 7 and rule of 21: 7 points above or below the mean indicate a run or a trend. 21 data points are required for a sample to be statistically valid. Control charts are used to identify special or assignable causes and to determine if a process is in control or out of control, The control chart includes the mean or center line and upper and lower control limits. UCL and LCL.

Pareto Diagram A Histogram that displays problems or issues by frequency of occurrence and is associated with the 80/20 rule. 80% of problems are associated with 20% of the causes.

Cost of Conformance Includes two components –Prevention (training, design reviews, and Appraisal (inspections).

Cost of Nonconformance Includes two components - Internal Failure (defects, rework) and , External Failure – customer receives a defective product.

Remember: DTRTRTFT Do the right thing right the first time.

ISO 9000 International certification indicating documented processes for quality. ISO 9001 is the most comprehensive certification in the series. ISO 9000 is updated approximately every four years.

Grade vs. Quality Low-grade products can be considered to be of high quality but may not posses the technical characteristics of other similar products. Low grade is not necessarily a problem but poor quality is always a problem. Example—software with limited functions may serve a specific need and is produced with high-quality, but limited features. (It does what it was intended to do, but it is low grade, that is, it does not have all of the features and functions of a higher grade product..

Quality The basic definition is "The conformance to specifications and fitness for use."

Essential Quality Characteristics Utility – ease of use, Financial— effective management of all associated costs, Legal – warranties,, safe for use, maintainability, reliability, availability..

Customer Satisfaction Managing to meet customer expectations. Define requirements and expectations early in the project and then review on a regular basis. Get to the desired result as expressed by the customer.

Prevention Over Inspection The cost of prevention is generally less than the activities of appraisal, inspection, and failure. Inspection adds more cost.

Management Responsibility Participation of all members is required to achieve high quality. According to Deming, management is responsible to provide the resources and is responsible for 85% of quality issues.

Shewhart Chart—Plan, Do, Check, Act The basic principles of continuous improvement. This process is performed in each project phase. (Deming cycle).. It is the basis for the 5 major processes – Initiating, Planning, Executing, Monitoring and Controlling, and closing

Attribute Sampling Pass or fail. Either reject or accept the unit being reviewed.

Variable Sampling Units are acceptable within a tolerance level and measured using scale to determine level of conformance..

Quality Control Measuring performance to determine where variances exist. Includes: Inspection, Control Charts, Pareto Diagrams (80/20 rule).

Statistical Sampling Single sampling, double sampling and multiple sampling. Producer's or Alpha risk – risk to a seller that a good lot will be rejected, Consumer's or Beta Risk – risk that a consumer will accept a bad lot. .

Six Sigma 3.4 defects per million opportunities. High cost during early part of implementation. Costs are recovered over time. The goal of Six Sigma— minimize defects through continuous analysis and action. The focus is toward removing defects. It includes the need for a positive attitude toward achieving the goals. Six Sigma is achieved through a process using tools such as Pareto charts, control charts, variance testing, Design of Experiments.

> **Sigma** Statistical expression indicating how much variation there is in a product.

> **Defect** Anything that will cause customer dissatisfaction.

> Six Sigma tools and processes eliminate unnecessary tasks and reduce or remove rework.
> 5 Sigma—230 defects per million opportunities
> 4 Sigma—6,210 defects per million opportunities
> 3 Sigma—66,800 defects per million opportunities
> 3 and 4 Sigma are associated with an average quality company

> **DMAIC** Define, Measure, Analyze, Improve, Control. DMAIC is associated with the processes found within Six Sigma

PROJECT HUMAN RESOURCES MANAGEMENT

Human Resources Management includes the processes that are used to organize and manage the project team. This includes selection of the project team, motivation and leadership, conflict management, the difference between authority and power, and performance appraisals. Organizational structure is generally included in Human Resources Management. Organizational structure refers to three basic types – Functional, Matrix, and Projectized structures.

Develop Human Resource Plan   Identifying project roles, establishing reporting relationships, and creating the staffing plan.

> **Inputs to Develop Human Resource Plan**   Activity resource requirements, enterprise environmental factors, organizational process assets
>
> **Key terms**   Organization chart, networking, organizational theory, roles and responsibilities

Acquire the Project Team   Obtaining the human resources needed to complete the project. This can be done through hiring, outsourcing, or obtaining internal resources. The organizational process assets will have an impact on the acquisition of the project team.

> **Key terms**   Pre-assignment - the team is selected and in place before the project manager is assigned, negotiation, acquisition – outsourcing or obtaining external resources, virtual teams (geographically dispersed teams)

Develop the Project Team   Building the project team, improving performance, improving teamwork

> **Key terms**   General management skills, interpersonal skills, training, team building activities, team ground rules, co-location, recognition, and reward

Manage the Project Team   Tracking the performance of the team and providing feedback on results, resolving problems and issues, coordinating changes, managing the effectiveness of the team to improve overall project performance.

> **Key terms**   Team performance assessment, project performance appraisals, observation and conversation, conflict management, issue log

Roles and Responsibilities   Should be identified and assigned during the project kickoff meeting.

RAM—Responsibility Assignment Matrix   The RAM is aligned with the WBS. The RAM links project team members with WBS tasks. The RAM is used to clearly define responsibility for the completion of project tasks. It is not a tool for determining resource requirements.

Interrelationships   The key interfaces a project manager will encounter and work with during the project. Examples—engineering, design, operations. Project Interfaces include Organizational (business units), Technical (design group and production group), and Interpersonal Relationships (formal and informal personal relationships).

Organizational Structure   Three major types of organizational structure -Functional, Matrix, and Projectized. Each typo of structure has unique characteristics. The authority of the project manager is lowest in the function structure, moderate in the matrix, and nearly total in the projectized structure.

Power and Authority There are five major types of power or influence:
Legitimate – based on formal position
Expert – based on knowledge and experience
Penalty – based on the ability and authority to penalize employees. Also known as coercive power
Reward – based on the ability and authority to offer items of perceived value to employees in exchange for work
Referent – influencing people through the power and authority of another

Conflict Management There are five general types of conflict handling: Withdrawing, Smoothing, Compromise, Forcing, Collaboration - through Confrontation and Problem Solving).

Major Sources of Conflict Schedules, different priorities, costs, estimates,, different opinions, lack of communication. Consider projects that you have worked on and identify other sources of conflict and the solutions to those conflicts. Escalation to the project sponsor or executive may be necessary to resolve some conflicts. Generally conflicts are resolved by understanding the needs and wants of the opposing person or group and identifying common areas of agreement.

Motivation Theories Theory X – an authoritative approach, Theory Y – a participative management style, and Z – Introduced by Ouchi- a communal type of structure., Maslow's Hierarchy of Needs, Herzberg -Hygienic factors and motivational factors

Motivation Douglas McGregor: Theory X—a micro-manager distrusts employees, believes employees do not want to work, and will do only what is minimally required. Significant levels of supervision are required Theory Y—Participative style, trusting, supportive. The manager believes people sincerely want to work and make a contribution. Theory Z – Ouchi believed that a structure in which everyone supported the greater good of the organization and where long term employment and loyalty to the organization were emphasized were key to motivating employees.

Maslow—Hierarchy of Needs Expressed graphically as a pyramid . The lowest or first level is physiological needs. As each motivating factor is achieved it is no longer a motivator as the person now seeks the next level. The 5 major levels are: Physiological needs, Safety and Security, Social Needs, Esteem, and Self-Actualization.

Herzberg's Theory The manager must address hygienic factors such as working conditions, relationships, level of supervision, and compensation before attempting motivation . If these factors are not managed and provided at a satisfactory level they may cause pain, discomfort, or conflict. They are not motivators but must be managed to remove any potential dissatisfaction among employees.. Once the hygiene factors are satisfied, motivating factors can be introduced. Examples of motivators are personal growth, advancement, increased responsibility, challenging work and recognition.

Team Building The project manager must remember the importance of team building and the phases of team development: Forming, Storming, Norming, and Performing. Conflict begins during the forming of the team, it intensifies during the storming phase, begins to subside in the norming phase, and is minimized in the performing stage. The project manager is a team leader and should determine the best methods to enhance the performance of the team through team building activities, reward and recognition, training, appraisals, co-location, support, direction, feedback, and performance appraisals.

Leadership Skills Project managers provide coaching, facilitation, authoritative leadership, supportive leadership, vision, direction, and motivation..

Organizational Influences Remember the difference between a project based and a nonproject based organization. Project based – the organization managers all work as project work or manages projects for external customers. Non project based organizations do not have a formal project management methodology or do not normally engage in project type work.

Project Stakeholders Anyone involved in or directly impacted either positively or negatively as a result of the project.

Negotiating Project managers must develop effective negotiating skills. This includes understanding other viewpoints and needs, creating an environment for effective negotiation and developing negotiation skills and tactics.. Items that may require negotiation include: obtaining resources, schedule development, activity duration estimates, funding requirements, resolving conflicts.

PROJECT COMMUNICATIONS MANAGEMENT

Project communications is about exchanging information and establishing and managing an effective flow of information between the project manager and stakeholders. Feedback about information provided is essential to effective communications. Effective communication includes the timely generation, collection, distribution, storage, and retrieval of project information.

Project communications planning is an essential part of the overall integrated project plan. It includes the processes that are required to be followed by the project team to effectively generate, collect, store, analyze, and distribute project-related information. Effective communication to project stakeholders is a key item in the successful achievement of project objectives. There are five major subprocesses associated with project communications: Identify Stakeholders, Plan Communications, Distribute Information, Manage Stakeholder Expectations, Report Performance

Identify Stakeholders It is important to identify everyone who will be involved in the project or affected by the project. Stakeholder identification also includes determining interests, levels of influence, and their definition of success and value.

> **Inputs to Identify Stakeholders** Project charter, procurement documents, organizational process assets, enterprise environmental factors
>
> **Key Terms** Stakeholder analysis, expert judgment, stakeholder register

Plan Communications Preparing for the information needs of the project stakeholders.

> **Inputs to Plan Communications** Stakeholder register, stakeholder management strategy, enterprise environmental factors, organizational process assets
>
> **Key terms** Project Scope Statement, Constraints, Assumptions, Communications Requirements Analysis, Communications Technology, communications models – sender –receiver model, Communications methods – interactive, push communications, pull communications. Communications Requirements relates to the specific needs of the stakeholders in terms of type of information and format (depends on the responsibility of the stakeholder). Communications Technology is associated with how information will be transferred to stakeholders in terms of information systems and capabilities of the stakeholders to access and send information.

Distribute Information Providing relevant project information to the stakeholders in a timely manner.

> **Inputs to Distribute Information** project management plan, performance reports, organizational process assets.

Key terms Communications skills, sender-receiver models, choice of media, writing style, meeting management techniques, presentation techniques, facilitation techniques information gathering and retrieval systems, information distribution methods, lessons learned process

Manage Stakeholder Expectations Ensuring that the needs of the identified stakeholders are met. Communicating with stakeholders to review and verify expectations .Understanding stakeholder requirements, managing project communications, and resolving issues between stakeholders.

Inputs to Manage Stakeholder Expectations Stakeholder register, Stakeholder management strategy, project management plan, issues log, change log, organizational process assets

Key terms Communications Methods, Interpersonal skills, Management skills – public speaking, writing, presentation, negotiation skills, Issue Logs, building trust, resolving conflict, active listening, managing resistance to change

Project managers communicate primarily in a horizontal or cross-organizational mode but may also communicate upward to executives and downward to employees and project performers. .

Report Performance Collecting, analyzing, and reporting project performance information to project stakeholders.

Inputs Project management plan, work performance information, work performance measurements, budget forecasts, organizational process assets

Key terms Variance analysis, forecasting methods, communications methods, reporting systems, deliverables, status review meetings, time reporting systems, cost reporting systems, performance reports, project performance forecasts.

Performance reporting is associated with collecting, reviewing, analyzing, and then disseminating the information to the appropriate stakeholders. Performance reporting includes status reports – the current project condition, progress reports – what has been accomplished to date, , and forecasts – what is expected or anticipated to occur in the future.

Earned Value Analysis is also associated with performance reporting—it utilizes three key values to assess project performance: Planned Value, or PV (also known as BCWS or Budgeted Cost of Work Scheduled), Actual Cost, or AC (also known as ACWP or Actual Cost of Work Performed), and Earned Value, or EV (also known as BCWP or Budgeted Cost of Work Performed). These values are used to calculate Cost Variance (EV – AC), Schedule Variance (EV – PV), Cost Performance Index (EV/AC), and Schedule Performance Index (EV/PV).

Other important Communications Management Terms and Study Items

Communication includes: internal communication – within the performing organization, external communication – with suppliers, the media, the public, formal and informal communications, vertical – upward and downward communications, horizontal communications – across organization's official and unofficial communications, verbal and non-verbal communications (body language)

Constraints Identified limitations that may obstruct or impede communications.

Assumptions Planning items provided to the team by the customer, sponsor or executive manager that are considered to be true, real, or certain.

Sender–Receiver Model basic elements – sender or transmitter, message, encode message, personality screen, sender region of experience, receiver, receiver perception screen, noise, medium (method to convey the message) receiver region of experience, receiver decode of message, feedback message, receiver personality screen, sender perception screen

Project communications is successfully achieved through an understanding of the sender–receiver model. In the basic communications model the critical elements are the region of experience of the sender and receiver, the message, encoding of the message, personality and perception screens, decoding and message received. An overlap he regions of experience or a similar backgrounds between sender and receiver will facilitate the communications process. The sender prepares or encodes a message to transmit to a receiver. The message passes through the sender's personality screen, across the region of experience to the receiver. The message passes through the receiver's perception screen and is then decoded. To ensure that the message was received as intended the receiver will feedback the message by encoding the message, passing it through the receiver's personality screen, through to region of experience, then through the sender's perception screen. The message is decoded by the original sender to determine if the message was sent and received as intended. The model illustrates the potential for distortion of communicated messages to occur due to common communications barriers associated with the personality and perception screens.

Communications Channel Formula: Determines the number of communications channels present within a project team. Knowledge of communications channels assists in developing plans to ensure effective communication and minimize potential breakdowns in communications.

Communications Channels (X) = N (N – 1)/2 where **N** = number of people on the project team.

Increasing the number of people on a project team increases the number of channels of communication that will exist between project team members and increases the potential for communications issues to develop..

PROJECT RISK MANAGEMENT

Project Risk Management includes the processes of risk management planning, identification of risks, prioritization of risks through analysis, responding to risks and controlling risks through the life cycle of the project.

Risk management is justifiable regardless of project size or complexity and should begin at the project kick off. Risk management should be included in each phase of the project and is an ongoing process.

There are six major processes in Project Risk Management:

Plan Risk Management   Deciding on the approach to conducting risk management and managing project risk.

Inputs to Plan Risk Management   Project scope statement, cost management plan, schedule management plan, communications management plan, enterprise environmental factors, organizational process assets.

Key terms   Planning meetings and analysis

Identify Risk   Determining which risks might affect the project and documenting their characteristics.

Inputs to Identify Risk   Risk management plan, activity cost estimates, activity duration estimates, scope baseline, stakeholder register, cost management plan, schedule management plan, quality management plan, , project documents, enterprise environmental factors, organizational process assets

Key terms Documentation reviews, information gathering techniques (Brainstorming, Delphi Technique, SWOT Analysis), Checklist Analysis, Assumptions Analysis, diagramming techniques, Risk Register, expert judgment

Perform Qualitative Risk Analysis Prioritizing risks for additional analysis or action by considering the probability and impact of the risk.

Inputs to Perform Qualitative Analysis Risk register, risk management plan, project scope statement, organizational process assets.

Key terms Probability and Impact Assessment, Probability and Impact Matrix, Risk data quality assessment, Risk Categorization, Risk Urgency Assessment, expert judgment

Perform Quantitative Risk Analysis A numerically based approach using mathematical models and simulations to determine the affect of risks.

Inputs to Quantitative Analysis Risk register, risk management plan, cost management plan, schedule management plan, organizational process assets.

Key terms Data gathering and representation techniques, interviewing, probability distributions—Normal, Beta, and Triangular distribution, , expert judgment, modeling techniques—sensitivity analysis, expected monetary value analysis, decision tree analysis, modeling and simulation—Monte Carlo Analysis)

Plan Risk Responses Determining the appropriate actions and identifying the options that may be used to reduce or eliminate project risks.

Inputs to Plan Risk Responses Risk register, risk management plan

Key terms Strategies for Negative Risks or Threats—Avoid, Transfer, Mitigate. Strategies for Positive Risks or Opportunities—Exploit, Share, Enhance. Strategies for both negative and positive risks—Acceptance (passive or active) and Contingency Response Strategy.

Monitor and Control Risk Tracking risks, identifying new risks, executing the risk response strategies, maintaining awareness of risks and their impact through the project life cycle.

Inputs to Monitor and Control Risk Risk register, project management plan, work performance information, performance reports

Key terms Risk reassessment, Audits, Risk reviews, Earned Value Analysis, Variance and Trend Analysis, Technical Performance Measurement, Reserve Analysis, Status Meetings

There are two primary components of risk—probability and impact. Urgency is also a factor when planning for risk and determining responses.

Other Important Risk management Terms and Study Items

Risk Tolerance There are risk seekers, risk averse organizations, and risk neutral organizations. Remember the Utility Factor – a measure of the tolerance for risk: Utility rises at a decreasing rate for the risk averter (the greater the risk the less the risk averter likes it). Utility increases for the risk seeker as risk intensity increases. Risk seekers thrive on risk opportunities. The utility factor rises at a uniform rate with increasing risk

Expected Monetary Value The product of probability and impact. Usually associated with Decision Trees.

Hurwitz and Wald Criterion (Maximax and Maximin) The Hurwitz criterion or maximax decision process is associated with organizations who have a high tolerance for risk and a "go-for-broke" attitude. The Wald or maximin criterion is associated with risk averse organizations who are concerned with loss.

Decision Tree Associated with Quantitative Risk Analysis. The decision tree assists in determining the potential outcomes of specific decisions by considering probability, initial cost, net path value and expected monetary value of each path. Decision trees assist in identifying the implications of choosing each of the available alternatives.

Brainstorming Gathering risk data from the project team by openly generating ideas and avoiding assessment of the ideas until the generation of new ideas has been exhausted.

Delphi Technique A process used in decision making and analysis where subject matter experts are engaged to provide input in an anonymous setting to eliminate bias.

Nominal Group Technique Similar to Delphi but done within group. The information gathered is anonymous through notes and written input but the participants are known to all.

Risk Triggers Symptoms that may lead to a risk event. It is important to address risk triggers before a serious risk event occurs.

Probability/Impact Matrix a matrix utilized to establish the value of a risk or establish a risk rating using the product of probability and impact. Scales may by ordinal - .1.2.3.4.5 or cardinal scales - low, moderate, high ratings.

Risk Response Strategies Avoidance, Transference, Mitigation, Exploit, Enhance, Share, Acceptance – passive acceptance means no specific action will be taken, active acceptance means a specific contingency will be planned.

Types of Risks Insurable Risk: chance for loss only. Business Risk: chance for profit or loss.

Monte Carlo Process A computer simulation of a project in which variable data is used for project activities, The project is simulated through software by introducing different possible activity outcomes that will affect the total project, Generally this type of simulation (similar to the throwing of dice) is done many times to determine possible outcomes and the probability achieving of task completions by a certain date or meeting project budget requirements.

PROCUREMENT MANAGEMENT

Project procurement includes the processes of determining what goods or services must be procured externally from an organization, make or buy decisions, buyer and seller relationships, types of contracts, and the risks associated with contract types. It also includes managing the seller, assign performance, managing contract terms, and closing contracts and procurement activities.

Project procurement management includes four major processes: Plan procurements, Conduct procurements, Administer procurements, Close procurements

Plan Procurements Determining what, when, and how to purchase or acquire products or services.

> **Inputs to Conduct Procurements** Scope baseline, requirements documentation, teaming agreements, risk register,, risk related contract decisions, activity resource requirements, project schedule activity cost estimates, cost performance baseline, enterprise environmental factors, organizational process assets.

> **Key terms** Make or Buy Analysis, Expert Judgment, Contract Types, Procurement Plan, Contract Statement of Work

Conduct Procurements The processes required to identify and document procurement requirements and determine potential suppliers or sellers.

> **Inputs to Conduct Procurements** Project management plan, procurement documents, project documents, source selection criteria, qualified seller list, seller proposals, make or buy decisions, teaming agreements, organizational process assets.

> **Key terms** Bidder conference, proposal evaluation techniques, independent estimates, expert judgment, advertising, internet search, procurement negotiations. Standard Forms, Procurement Documents, Evaluation Criteria

> **Request Seller Responses** This process is now included under Conduct procurements in the *PMBOK® Guide*—Fourth edition. It includes preparing requests for information, obtaining information, quotes, bids, offers, proposals from potential suppliers and sellers

> **Select Sellers** This process is now included under Conduct Procurements in the *PMBOK® Guide*—Fourth edition. Select sellers includes determining which sellers or providers are qualified to perform the required services and developing the appropriate contracts

> **Key terms** Weighting System, Independent Estimates, Screening System, Contract Negotiation, Seller Rating Systems, Expert Judgment, Proposal Evaluation Techniques, Contract, Contract Management Plan

Administer Procurements Managing the contract terms and conditions and the procurement relationships between buyer and seller. Managing changes and documenting the performance of the seller during execution.

> **Inputs to Administer Procurements** Procurement documents, project ,management plan,, contract, performance reports, approved change requests, work performance information,

> **Key terms** Contract Change Control System, Procurement performance Review, Inspections and Audits, Performance Reporting, Payment System, Claims Administration, Records Management System, Information Technology

Close Procurements Completing all procurement work and settling the contract, resolving open issues and closing all contractual items.

> **Inputs to Close Procurements** Project management plan, procurement documentation

> **Key terms** Procurement Audits, negotiated settlements, Closed Contracts, Contract File, Deliverable Acceptance, Lessons Learned Documentation

Study Items:

Remember to review the different types of contracts and the advantages and disadvantages of each type (Lump sum vs. Cost Plus Contracts). In a lump sum or Firm Fixed Price Contract (FFP) the risk is associated with the seller. In Cost Plus Contracts the risk is associated with the buyer. Review the common types of contracts: Firm Fixed Price or Lump Sum, Fixed Prices with Incentive Fee (FPIF) Cost Plus Incentive Fee (CPIF), Cost Plus Percentage of Cost (CPPC). Also remember to review contracts with sharing ratios. These are associated with incentive type contracts where the buyer and seller share realized cost reductions or meet project deliverables at a lower cost than originally planned.

Contract A Legal document that includes an offer, acceptance, and consideration (something of value).

> Definitive Contract   The final, agreed upon contract

> Completion Contract   The contractor/ supplier/ seller is required to produce a specific end product or deliverable

> Term Contract A contract that is intended to deliver a specific level of effort, not an actual product.

Statement of Work A narrative description of work to be completed under contract. This document provides details about the work to be done by the contractor.

Make or Buy Decisions Make or buy decisions depend on cost, desired level of control, capability of the buyer's organization, type of deliverable, capacity of the buyer's organization. Consider benefits and disadvantages of the make and buy decisions and the risks that may be associated with the decision.

Bidder Conference A meeting scheduled by the buyer to provide information to all potential bidders. The meeting provides a level playing field for potential contractors.

Negotiation Negotiation is a key skill and necessary activity in the procurement process. Location for negations, the negotiation environment, needs of each party, hygiene issues (type of room set up, location, type of tables and positioning of the parties involved). Understanding the opposing viewpoint and using appropriate tactics.- Fair and reasonable offers, good faith,, protocol,

Special Terms and Conditions Make sure to identify Penalty clauses, liquidated damages clauses, force majeure clauses (natural disasters, events that cannot be controlled by the contractor) and other contract terms and conditions.

Force Majeure A common clause in contracts which essentially frees both parties from liability or obligation when an extraordinary event or circumstance beyond the control of the parties, such as a war, strike, riot, crime, or act of God (e.g., flooding, earthquake, volcano),

Privity   In contract law provides that a contract cannot confer rights or impose obligations arising under it on any person or agent except the parties to it.

Penalty Clause A provision in a contract that stipulates an excessive pecuniary charge against a defaulting party.

Liquidated Damages A contractually agreed upon amount to be paid in the event of a breach of the contract, in lieu of performance or quantification of actual damages sustained.

Contract Types Firm Fixed Price (FFP) or Lump Sum, Cost Plus Incentive Fee (CPIF), Fixed Price Incentive Fee (FPIF), Time and Material, Purchase Order, Cost Plus Percentage of Cost (CPPC), Cost Plus Fixed Fee (CPFF). Remember the risks associated with each type of contract and who is exposed to the greatest risk. Incentive contracts will include a sharing ratio.

Contract Close-out Development of Punch Lists, verification of contracted deliverables, formal acceptance, post project review, contract reviews.

Termination for Convenience A provision in the contract that allows the project to be terminated for other than performance reasons. Example—the technology is no longer needed, there are no additional funds to support the project.

Termination Due to Default A breach in the agreement, failure to provide what was agreed upon in the contract.

Non Disclosure agreement A **non-disclosure agreement** (NDA), also known as a **confidentiality agreement**, **confidential disclosure agreement** (CDA), **proprietary information agreement** (PIA), or **secrecy agreement**, is a legal contract between at least two parties that outlines confidential materials or knowledge the parties wish to share with one another for certain purposes, but wish to restrict access to. It is a contract through which the parties agree not to disclose information covered by the agreement

Fait accompli An accomplished fact; a thing already done

Administrative Changes Generally this type of change is for records only and does not impact the work of the project.

Constructive Change An action or inaction by a party associated with the project that impacts the actual work and requires a change to the planned work.

Project Reviews At project completion the project manager and team conduct reviews of the entire project and compare actual results with contractual agreements.

Close-out Punch lists are prepared, verification of contracted deliverables, formal acceptance is obtained, post project reviews are conducted, contract reviews are conducted, final payments are made.

Professional Responsibility

There are four elements in the domain known as Professional Responsibility:
Ensuring Individual Integrity, Contributing to the Project Management Knowledge Base, Enhancing Personal and Professional Competence, and Promoting Interaction Among Stakeholders.

Professional Responsibility is closely associated with the *PMBOK® Guide* Knowledge areas: Human Resource Management and Communications Management. Professional Responsibility is also associated with diversity, cultural differences, time zones, geographic location, political issues, ethics, ethnic differences and customs. PMP® candidates should include in their study the issues and sensitivities of dealing in an international project environment.

Professional responsibility includes the following subjects:

Ensuring Individual Integrity
Responsibilities of employers
Accommodating employee needs

Diversity in the workplace
Complying with federal employment laws
Dealing ethically with project stakeholders
Making ethical business decisions (Honesty and integrity)
Dealing ethically with employees, customers, and other businesses

Promoting Interaction Among Stakeholders
Listening skills
Communication styles (authoritative, facilitating, judicial)
Effective negotiation and conflict resolution (balancing stakeholder needs)
Recognizing conflict and managing conflicts effectively
Conflict management—Collaboration, Compromise, Forcing, Withdrawing, Smoothing

Enhancing Personal and Professional Competence
Creating a personal development plan
Mentor roles and skills

Contributing to the Project Management Knowledge Base
Documenting and sharing lessons learned

Project Management Formulas

The following is a summary list of formulas that may be associated with PMP® exam questions:

Communications channels $X = N9N-1) /2$

Weighted average formula (beta distribution) (Optimistic + 4 times the most likely value + pessimistic) / 6

Float – obtained during the forward and backward pass calculations in CPM. Float is the diffeence between the late finish and early finish of an activity (check the logical relationships before determining float)

Standard Deviation   (Pessimistic – Optimistic) / 6

Net Present Value   The sum of the present values minus the initial cost

Present Value $FV / (1+r)^n$

Future Value $PV (1 +r)^n$

Earned Value Formulas

EV (BCWP) – AC (ACWP) = Cost variance (CV) (a negative cost variance = budget overrun)
EV – PV (BCWS) = Schedule variance (SV) (a negative SV means the project or work package is behind schedule)
EV/AC (ACWP) = Cost performance index (CPI) (a measure of efficiency in managing the project budget)
EV/PV (BCWS) = Schedule performance index (SPI) (a measure of efficiency in managing the project schedule)

A performance index (PI) that equals 1 means there is no variance and the project or work package is is performing as planned. A PI greater than 1 indicates superior performance. A PI less than 1 indicates a a schedule slippage or a cost overrun.

EAC Estimate at Complete $EAC = AC^C + ETC$ (Estimate to Complete) – this formula is used when new estimates are developed for remaining work.

$EAC = AC^C + BAC - EV^C$ where BAC is the Budget at Complete   This formula is used when variances are seen as atypical and similar variances are not anticipated to occur again.

$EAC = AC^C + (BAC - EV^C / CPI^C$ Used when current variances are seen as typical and will be seen in the future

Your Personal Learning Library

Write down your thoughts, ideas, and observations about the material in the chapter that may assist you with your learning experience. Create action items and additional study plans to assist you in enhancing your skills or for preparing to take the PMP® or CAPM® exam.

Insights, key learning points, personal recommendations for additional study, areas for review, application to your work environment, items for further discussion with associates.

_____

_____

_____

_____

_____

_____

_____

_____

Personal Action Items:	
Action Item	**Target Date for Completion**

Practice Questions for the Project Management Professional—PMP® Exam

The following multiple-choice questions will be helpful in reviewing **Integration Management:**

1. The triple constraints on a project are:
   a. Time, cost, and profitability
   b. Resources required, sponsorship involvement, and funding
   c. Time, cost, and quality and/or scope
   d. Calendar dates, facilities available, and funding

2. Which of the following is not part of the definition of a project?
   a. Repetitive activities
   b. Constraints
   c. Consumption of resources
   d. A well-defined objective

3. Which of the following is usually not part of the criteria for project success?
   a. Customer satisfaction
   b. Customer acceptance
   c. Meeting at least 75 percent of specification requirements
   d. Meeting the triple constraint requirements

4. Which of the following is generally not a benefit achieved from using project management?
   a. Flexibility in the project's end date
   b. Improved risk management
   c. Improved estimating
   d. Tracking of projects

5. To whom would the project manager go first for assigning the resources to a project?
   a. The project management
   b. The Human Resources Department
   c. The line manager
   d. The executive sponsor

6. Where would the project manager go first for resolving conflicts between the project and line managers?
   a. Assistant project manager for conflicts
   b. The project sponsor
   c. The executive steering committee
   d. The Human Resources Department

7. After the project is initiated, the customer wants some degree of confidence that the customer's objectives are achievable. Which document would the project manager use first to convince the customer?
   a. The customer's statement of work
   b. The project manager's scope statement
   c. The project plan
   d. The scope baseline

8. The project manager normally meets with the project sponsor during project initiation to get assistance in:
   a. Defining the project's objectives in both business and technical terms
   b. Developing the project plan
   c. Performing the project feasibility study
   d. Performing the project cost-benefit analysis

9. The role of the project sponsor during project execution is to:
   a. Validate the project's objectives
   b. Validate the execution of the plan
   c. Make all project decisions
   d. Resolve problems/conflicts that cannot be resolved elsewhere in the organization

10. The role of the project sponsor during the closure of the project or a life-cycle phase of the project is to:
   a. Validate that the profit margins are correct
   b. Sign off on the acceptance of the deliverables
   c. Administer performance reviews of the project team members
   d. Get the customer to agree to follow-on work

Answers

1. c
2. a
3. c
4. a
5. c
6. b
7. c
8. a
9. d
10. b

The following multiple-choice questions will be helpful in reviewing the principles of **Project Scope Management:**

1. Which document would the project manager identify first to the project team at the initial kickoff meeting to show that the project is officially sanctioned as a project?
   a. Project charter
   b. Project plan
   c. Feasibility study
   d. Cost-benefit analysis

2. A project manager is unsure of what is required in a certain work package. To get a better understanding of what the work package involves, the project manager would first look at the:
   a. Schedule control documents
   b. Code of accounts
   c. WBS dictionary
   d. Risk management plan

Answer Questions 3–6 using the work breakdown structure shown below (numbers in parentheses show the dollar value for a particular element):

```
1.00.00
    1.1.0           ($25k)
        1.1.1
        1.1.2       ($12k)
    1.2.0
        1.2.1       ($16k)
        1.2.2.0
            1.2.2.1   ($20k)
            1.2.2.2   ($30k)
```

3. The cost of WBS element 1.2.2.0 is:
   a. $20K
   b. $30K
   c. $50K
   d. Cannot be determined

4. The cost of WBS element 1.1.1 is:
   a. $12K
   b. $13K
   c. $25K
   d. Cannot be determined

5. The cost of the entire program (1.00.00) is:
   a. $25K
   b. $66K
   c. $91K
   d. Cannot be determined

6. The work packages in the WBS are at WBS level(s):
   a. 2 only
   b. 3 only
   c. 4 only
   d. 3 and 4

7. One of the outputs of the PMBOK® Scope Definition Process is:
   a. A project charter
   b. A scope statement
   c. A detailed WBS
   d. None of the above

8. One of your contractors has sent you an e-mail requesting that they be allowed to conduct only eight tests rather than the ten tests required by the specification. What should the project manager do first?
   a. Change the scope baseline
   b. Ask contractor to put forth a change request
   c. Look at the penalty clauses in the contract
   d. Ask your sponsor for his/her opinion

9. One of your contractors sends you an e-mail request to use high-quality raw materials in your project stating that this will be value-added and improve quality. What should the project manager do first?
   a. Change the scope baseline
   b. Ask the contractor to put forth a change request

    c. Ask your sponsor for his/her opinion

    d. Change the WBS

10. The change control board, of which you are a member, approves a significant scope change. The first document that the project manager should update would be the:

    a. Scope baseline

    b. Schedule

    c. WBS

    d. Budget

Answers to multiple-choice questions:

  1. a

  2. c

  3. c

  4. b

  5. c

  6. d

  7. b

  8. b

  9. b

10. a

The following multiple-choice questions will be helpful in reviewing the principles of **Project Time Management:**

1. The shortest time necessary to complete all of the activities in a network is called the:

    a. Activity duration length

    b. Critical path

    c. Maximum slack path

    d. Compression path

2. Which of the following *cannot* be identified after performing a forward and backward pass?

    a. Free float

    b. Slack time

    c. Critical path activities

    d. How much overtime is planned

3. Which of the following is *not* a commonly used technique for schedule compression?

    a. Resource reduction

    b. Reducing scope

    c. Fast-tracking activities

    d. Use of overtime

4. A network-based schedule has four paths, namely 7, 8, 9, and 10 weeks. If the 10-week path is compressed to 8 weeks, then:

    a. We now have two critical paths

    b. The 9-week path is now the critical path

    c. Only the 7-week path has slack

    d. Not enough information is provided to make a determination

5. The major disadvantage of using bar charts to manage a project is that bar charts:

    a. Do not show dependencies between activities

    b. Are ineffective for projects under one year in length
    c. Are ineffective for projects under $1 million in size
    d. Do not identify start and end dates of a schedule

6. Which of the following steps would a project do first in the development of a network diagram?
    a. Listing of the activities
    b. Determination of dependencies
    c. Calculation of effort
    d. Calculation of durations

7. Reducing the peaks and valleys in manpower assignments in order to obtain a relatively smooth manpower curve is called:
    a. Manpower allocation
    b. Resource leveling
    c. Resource allocation
    d. Resource commitment planning

8. Activities with no time duration are called _____ activities.
    a. Reserve
    b. Dummy
    c. Zero slack
    d. Supervision

9. Optimistic, pessimistic, and most likely activity times are associated with:
    a. PERT
    b. GERT
    c. PDM
    d. ADM

10. The most common "constraint" or relationship in a precedence network is:
    a. Start-to-start
    b. Start-to-finish
    c. Finish-to-start
    d. Finish-to-finish

Answers to multiple-choice questions:

1. b
2. d
3. a
4. d
5. a
6. a
7. b
8. b
9. a
10. c

The following multiple-choice questions will be helpful in reviewing the principles of **Project Cost Management:**

1. In earned value measurement, earned value is represented by:
   a. BCWS
   b. BCWP
   c. ACWP
   d. None of the above

2. If BCWS = 1,000, BCWP = 1,200, and ACWP = 1,300, the project is:
   a. Ahead of schedule and under budget
   b. Ahead of schedule and over budget
   c. Behind schedule and over budget
   d. Behind schedule and under budget

3. If BAC = $20,000 and the project is 40% complete, then the earned value is:
   a. $5,000
   b. $8,000
   c. $20,000
   d. Cannot be determined

4. If BAC = $12,000 and CPI = 1.2, then the variance at completion is:
   a. −$2,000
   b. +$2,000
   c. −$3,000
   d. +$3,000

5. If BAC = $12,000 and CPI = 0.8, then the variance at completion is:
   a. −$2,000
   b. +$2,000
   c. −$3,000
   d. +$3,000

6. If BAC for a work package is $10,000 and BCWP = $4,000, then the work package is _____ percent complete.
   a. 40
   b. 80
   c. 100
   d. 120

7. If CPI = 1.1 and SPI = 0.95, then the trend for the project is:
   a. Running over budget but ahead of schedule
   b. Running over budget but behind schedule
   c. Running under budget but ahead of schedule
   d. Running under budget but behind schedule

8. The document that describes a work package, identifies the cost centers allowed to charge against this work package, and establishes the charge number for this work package is the:
   a. Code of accounts
   b. Work breakdown structure
   c. Work authorization form
   d. None of the above

9. Unknown problems such as escalation factors are often budgeted for using the:
   a. Project manager's charge number
   b. Project sponsor's charge number
   c. Management reserve
   d. Configuration management cost account

10. EAC, ETC, SPI, and CPI most often appear in which type of report?
    a. Performance
    b. Status
    c. Forecast
    d. Exception

11. If BAC = $24,000, BCWP = $12,000, ACWP = $10,000, and CPI = 1.2, then the cost that remains to finish the project is:
    a. $10,000
    b. $12,000
    c. $14,000
    d. Cannot be determined

12. There are several purposes for the 50%–50% rule, but the *primary* purpose of the 50%–50% rule is to calculate:
    a. BCWS
    b. BCWP
    c. ACWP
    d. BAC

13. When a project is completed, which of the following *must* be true?
    a. BAC = ACWP
    b. ACWP = BCWP
    c. SV = 0
    d. BAC = ETC

14. In March, CV = –$20,000 and in April, CV = –$30,000. In order to determine whether or not the situation has really deteriorated because of a larger unfavorable cost variance, we would need to calculate:
    a. CV in percent
    b. SV in dollars
    c. SV in percent
    d. All of the above

15. If a project manager is looking for revenue for a value-added scope change, the project manager's first choice would be:
    a. Management reserve
    b. Customer-funded scope change
    c. Undistributed budget
    d. Retained profits

16. A project was originally scheduled for 20 months. If CPI is 1.25, then the new schedule date is:
    a. 16 months
    b. 20 months
    c. 25 months
    d. Cannot be determined

17. The cost or financial baseline of a project is composed of:
    a. Distributed budget only
    b. Distributed and undistributed budgets only
    c. Distributed budget, undistributed budget, and the management reserve only
    d. Distributed budget, undistributed budget, management reserve, and profit only

Answers to multiple-choice questions:

1. b
2. b
3. b
4. b
5. c
6. a
7. d
8. c
9. c
10. c
11. a
12. b
13. c
14. a
15. b
16. d
17. b

The following multiple-choice questions will be helpful in reviewing the principles of **Project Quality Management:**

1. Which of the following is not part of the generally accepted view of quality today?
   a. Defects should be highlighted and brought to the surface
   b. We can inspect in quality
   c. Improved quality saves money and increases business
   d. Quality is customer-focused

2. In today's view of quality, who defines quality?
   a. Contractor's senior management
   b. Project management
   c. Workers
   d. Customers

3. Which of the following are tools of quality control?
   a. Sampling tables
   b. Process charts
   c. Statistical and mathematical techniques
   d. All of the above

4. Which of the following is true of modern quality management?
   a. Quality is defined by the customer
   b. Quality has become a competitive weapon
   c. Quality is now an integral part of strategic planning
   d. All are true

5. A company dedicated to quality usually provides training for:
   a. Senior management and project managers
   b. Hourly workers
   c. Salaried workers
   d. All employees

6. Which of the following quality gurus believe "zero defects" is achievable?
   a. Deming
   b. Juran
   c. Crosby
   d. All of the above

7. What are the components of Juran's Trilogy?
   a. Quality Improvement, Quality Planning, and Quality Control
   b. Quality Improvement, Zero Defects, and Quality Control
   c. Quality Improvement, Quality Planning, and PERT Charting
   d. Quality Improvement, Quality Inspections, and Quality Control

8. Which of the following is not one of Crosby's Four Absolutes of Quality?
   a. Quality means conformance to requirements
   b. Quality comes from prevention
   c. Quality is measured by the cost of conformance
   d. Quality means that the performance standard is "zero defects"

9. According to Deming, what percentage of the costs of quality is generally attributable to management?
   a. 100%
   b. 85%
   c. 55%
   d. 15%

10. Inspection:
   a. Is an appropriate way to ensure quality
   b. Is expensive and time-consuming
   c. Reduces rework and overall costs
   d. Is always effective in stopping defective products from reaching the customer

11. The Taguchi Method philosophies concentrate on improving quality during the:
   a. Conceptual Phase
   b. Design Phase
   c. Implementation Phase
   d. Closure Phase

12. A well-written policy statement on quality will:
   a. Be a statement of how, not what or why
   b. Promote consistency throughout the organization and across projects
   c. Provide an explanation of how customers view quality in their own organizations
   d. Provide provisions for changing the policy only on a yearly basis

13. Quality assurance includes:
   a. Identifying objectives and standards
   b. Conducting quality audits
   c. Planning for continuous collection of data
   d. All of the above

14. What is the order of the four steps in Deming's Cycle for Continuous Improvement?
    a. Plan, do, check, and act
    b. Do, plan, act, and check
    c. Check, do, act, and plan
    d. Act, check, do, and plan

15. Quality audits:
    a. Are unnecessary if you do it right the first time
    b. Must be performed daily for each process
    c. Are expensive and therefore not worth doing
    d. Are necessary for validation that the quality policy is being followed and adhered to

16. Which of the following is/are typical tool(s) of statistical process control?
    a. Pareto analysis
    b. Cause-and-effect analysis
    c. Process control charts
    d. All of the above

17. Which of the following methods is best suited to identifying the "vital few"?
    a. Pareto analysis
    b. Cause-and-effect analysis
    c. Trend analysis
    d. Process control charts

18. When a process is set up optimally, the upper and lower specification limits typically are:
    a. Set equal to the upper and lower control limits
    b. Set outside the upper and lower control limits
    c. Set inside the upper and lower control limits
    d. Set an equal distance from the mean value

19. The upper and lower control limits are typically set:
    a. One standard deviation from the mean in each direction
    b. 3 σ (three sigma) from the mean in each direction
    c. Outside the upper and lower specification limits
    d. To detect and flag when a process may be out of control

20. Which of the following is *not* indicative of today's views of the quality management process applied to a given project?
    a. Defects should be highlighted and brought to the surface
    b. The ultimate responsibility for quality lies primarily with senior management or sponsor, but everyone should be involved
    c. Quality saves money
    d. Problem identification leads to cooperative solutions

21. If the values generated from a process are normally distributed around the mean value, what percentage of the data points generated by the process will *not* fall within +/– three standard deviations of the mean?
    a. 99.7%
    b. 95.4%
    c. 68.3%
    d. 0.3%

Answers to multiple-choice questions:

1. b
2. d
3. d
4. d
5. d
6. c
7. a
8. c
9. b
10. b
11. b
12. b
13. d
14. a
15. d
16. d
17. a
18. b
19. b
20. b
21. d

The following multiple-choice questions will be helpful in reviewing the principles of **Project Human Resources Management:**

1. In which organizational form is it most difficult to integrate project activities?
   a. Classical/traditional
   b. Projectized
   c. Strong matrix
   d. Weak matrix

2. In which organization form would the project manager possess the greatest amount of authority?
   a. Classical/traditional
   b. Projectized
   c. Strong matrix
   d. Weak matrix

3. In which organizational form does the project manager often have the least amount of authority?
   a. Classical/traditional
   b. Projectized
   c. Strong matrix
   d. Weak matrix

4. In which organizational form is the project manager least likely to share resources with other projects?
   a. Classical/traditional
   b. Projectized
   c. Strong matrix
   d. Weak matrix

5. In which organizational form do project managers have the greatest likelihood of possessing reward power and have a wage and salary administration function? (The project and line manager are the same person.)
   a. Classical/traditional
   b. Projectized
   c. Strong matrix
   d. Weak matrix

6. In which organizational form is the worker in the greatest jeopardy of losing his/her job if the project gets cancelled?
   a. Classical/traditional
   b. Projectized
   c. Strong matrix
   d. Weak matrix

7. In which type of matrix structure would a project manager most likely have a command of technology?
   a. Strong matrix
   b. Balanced matrix
   c. Weak matrix
   d. Cross-cultural matrix

8. Which of the following is not one of the sources of authority for a project manager?
   a. Project charter
   b. Job description for a project manager
   c. Delegation from senior management
   d. Delegation from subordinates

9. Which form of power do project managers that have a command of technology and are leading R&D projects most frequently use?
   a. Reward power
   b. Legitimate power
   c. Expert power
   d. Referent power

10. If a project manager possesses penalty (or coercive) power, he most likely also possesses:
    a. Reward power
    b. Legitimate power
    c. Expert power
    d. Referent power

11. A project manager with a history of success in meeting deliverables and in working with the team members would most likely possess a great deal of:
    a. Reward power
    b. Legitimate power
    c. Expert power
    d. Referent power

12. Most project managers are motivated by which level of Maslow's Hierarchy of Human Needs?
    a. Safety
    b. Socialization
    c. Self-esteem
    d. Self-actualization

13. You have been placed in charge of a project team. The majority of the team members have less than two years of experience working on project teams and most of the people have never worked with you previously. The leadership style you would most likely select would be:
    a. Telling
    b. Selling
    c. Participating
    d. Delegating

14. You have been placed in charge of a new project team and are fortunate to have been assigned the same people that worked for you on your last two projects. Both previous projects were very successful and the team performed as a high-performance team. The leadership style you would most likely use on the new project would be:
    a. Telling
    b. Selling
    c. Participating
    d. Delegating

15. Five people are in attendance in a meeting and are communicating with one another. How many two-way channels of communication are present?
    a. 4
    b. 5
    c. 10
    d. 20

16. A project manager provides a verbal set of instructions to two team members on how to perform a specific test. Without agreeing or disagreeing with the project manager, the two employees leave the project manager's office. Later, the project manager discovers that the tests were not conducted according to his instructions. The most probable cause of failure would be:
    a. Improper encoding
    b. Improper decoding
    c. Improper format for the message
    d. Lack of feedback on instructions

17. A project manager that allows workers to be actively involved with the project manager in making decisions would be using a _____ leadership style.
    a. Passive
    b. Participative/Democratic
    c. Autocratic
    d. Laissez-faire

18. A project manager that dictates all decisions and does not allow for any participation by the workers would be using a _____ leadership style.
    a. Passive
    b. Participative/Democratic
    c. Autocratic
    d. Laissez-faire

19. A project manager that allows the team to make virtually all of the decisions without any involvement by the project manager would be using a _____ leadership style.
    a. Passive
    b. Participative/Democratic
    c. Autocratic
    d. Laissez-faire

20. During project staffing, the *primary* role of senior management is in the selection of the _____.
   a. Project manager
   b. Assistant project managers
   c. Functional team
   d. Executives do not get involved in staffing

21. During project staffing, the *primary* role of line management is:
   a. Approving the selection of the project manager
   b. Approving the selection of the assistant project managers
   c. Assigning functional resources based upon who is available
   d. Assigning functional resources based upon availability and the skill set needed

22. A project manager is far more likely to succeed if it is obvious to everyone that:
   a. The project manager has a command of technology
   b. The project manager is at a higher pay grade that everyone else on the team
   c. The project manager is over 45 years of age
   d. Executive management has officially appointed the project manager

23. Most people believe that the best way to train someone in project management is through:
   a. On-the-job training
   b. University seminars
   c. Graduate degrees in project management
   d. Professional seminars and meetings

24. In staffing negotiations with the line manager, you identify a work package that requires a skill set of a Grade 7 worker. The line manager informs you that he will assign a Grade 6 and a Grade 8 worker. You should:
   a. Refuse to accept the Grade 6 because you are not responsible for training
   b. Ask for two different people
   c. Ask the sponsor to interfere
   d. Be happy! You have two workers.

25. You priced out a project work package at 1,000 hours assuming a Grade 7 employee would be assigned. The line manager assigns a Grade 9 employee. This will result in a significant cost overrun. The project manager should:
   a. Reschedule the start date of the project based upon the availability of a Grade 7
   b. Ask the sponsor for a higher priority for your project
   c. Reduce the scope of the project
   d. See if the Grade 9 can do the job in less time

26. As a project begins to wind down, the project manager should:
   a. Release all nonessential personnel so that they can be assigned to other projects
   b. Wait until the project is officially completed before releasing anyone
   c. Wait until the line manager officially requests that the people be released
   d. Talk to other project managers to see who wants your people

Answers:

1. a
2. b
3. d
4. b
5. b

6. b
7. a
8. d
9. c
10. a
11. d
12. d
13. a
14. d
15. c
16. d
17. b
18. c
19. d
20. a
21. d
22. d
23. a
24. d
25. d
26. a

The following multiple-choice questions will be helpful in reviewing the principles of **Project Communications Management:**

1. The sender–receiver model is designed to help project managers and team members improve their communications skills by emphasizing:
   a. The need for the project manager to engage in downward communication
   b. The importance of horizontal communications during project execution
   c. Potential communications barriers and the importance of feedback loops
   d. The need to communicate information verbally

2. To do communications planning, the process requires which of the following inputs?
   a. Work results, project plan, project records
   b. Communications requirements, communications technology, constraints, assumptions
   c. Performance reviews, trend analysis
   d. Project reports, product documentation, information retrieval systems

3. As a project manager you must identify the needs of the stakeholders involved in the project. These needs are identified in which of the following processes?
   a. Performance reviews
   b. Trend analysis
   c. Performance measurement
   d. Communications requirements

4. On most projects, the majority of communications planning is done
   a. During execution
   b. At the earliest stages of the project
   c. Just prior to closeout
   d. Evenly throughout each project phase

5. Project resources should be expended only on communicating information that:
   a. Contributes to the success of the project or where a lack of communication can lead to failure
   b. Comes from external sources such as government or regulatory agencies
   c. Has been generated internally by the organization's executive team
   d. Is developed by the project team

6. A factor in the determination of communications technology needs is:
   a. Stakeholder analysis
   b. The communications plan
   c. The immediacy of the need for information
   d. Communications skills of the project team and project manager

7. A collection and filing structure that details what methods will be used to gather and store various types of project information is the:
   a. Project management information system
   b. Communications management plan
   c. Scope statement
   d. Project assumption set

8. During project implementation, the project manager is required to submit several reports that will communicate information about the project. Describing where the project stands at the present moment in relationship to schedule and other metrics is found in the:
   a. Status report
   b. Progress report
   c. Project forecast
   d. Quality report

9. A commonly used method for determining and then reporting project performance that integrates scope, time, and cost for assessment of results is:
   a. Trend analysis
   b. Variance analysis
   c. Earned value analysis
   d. Project presentations and review

10. The cost performance index or CPI is used to indicate:
    a. The sum of individual earned value budgets
    b. Cost efficiency
    c. The variance between EV and PV
    d. The variance between actual cost and budgeted cost

11. When a project or phase has been completed the project team should document results and formalize acceptance through the process of:
    a. Change control
    b. Earned value analysis
    c. Administrative closure
    d. Performance reporting

12. Your team currently includes 9 members. You increase the team size to 12. The number of communications channels within the team has increased by:
    a. 10
    b. 30
    c. 9
    d. 54

Answers to multiple-choice questions:

1. c
2. b
3. d
4. b
5. a
6. c
7. b
8. a
9. c
10. b
11. c
12. b

The following multiple-choice questions will be helpful in reviewing the principles of **Project Risk Management:**

1. The two major components of a risk are:
   a. Time and cost
   b. Uncertainty and impact
   c. Quality and time
   d. Cost and decision-making circumstances

2. When setting up a system to identify or classify risks, the project manager would most likely classify the risks according to:
   a. The probability of occurrence
   b. The magnitude of the impact, favorable or unfavorable
   c. The source of the risk
   d. The customer's risk identification system

3. If there is a 40 percent chance of making $100,000 and a 60 percent chance of losing $150,000, then the expected monetary outcome is:
   a. $50,000
   b. –$50,000
   c. $90,000
   d. –$90,000

4. The project is managing a project that has a very large profit potential for the company. However, there exists the possibility that part of work can be outsourced in which case proprietary knowledge will have to be provided to the contractor. If the project manager is unwilling to release proprietary knowledge, the project manager would most likely avoid the risk response method of:
   a. Assumption
   b. Mitigation
   c. Avoidance
   d. Transfer

5. One technique for risk evaluation uses a questionnaire, a series of rounds, reports submitted in confidence and then circulated with the source unidentified. If the project manager uses this technique, he/she is using:
   a. The Delphi Technique
   b. The Work Group
   c. Unsolicited team responses
   d. A risk management team

6. In evaluating whether or not a risk is about to occur, the project manager would most likely look at which of the following first?
   a. Budgets
   b. Triggers
   c. Schedules
   d. Project plan

7. The project manager needs a tool for assessing or quantifying a risk. The project manager would accept all of the following except:
   a. Decision Tree analysis
   b. Objective setting
   c. Simulation
   d. Interviewing

8. The project manager realizes that there is a high probability that risks will occur on the project, and some technique is necessary. A technique that depicts interactions among decisions and associated events is needed. The project manager would most likely look at which of the following first?
   a. Decision Tree analysis
   b. Earned value measurement system
   c. Network scheduling system
   d. Payoff matrix

9. Which risk response strategy would the project manager do first that reduces the probability or impact of the event without altering the project's objectives?
   a. Avoidance
   b. Acceptance
   c. Mitigation
   d. Transfer

10. Earned value measurement would be an example of:
   a. Risk communication planning
   b. Risk assessment
   c. Risk response
   d. Risk monitoring and control

Answers to multiple-choice questions:

1. a
2. c
3. b
4. d
5. a
6. b
7. b
8. a
9. c
10. d

The following multiple-choice questions will be helpful in reviewing the principles of **Project Procurement Management:**

1. The contractual Statement of Work document is:
   a. A nonbinding legal document used to identify the responsibilities of the contractor
   b. A definition of the contracted work for government contracts only
   c. A narrative description of the work/deliverables to be accomplished and/or the resource skills required
   d. A form of specification

2. A written or pictorial document that describes, defines, or specifies the services or items to be procured is:
   a. A specification document
   b. A Gantt chart
   c. A blueprint
   d. A risk management plan

3. The "Order of Precedence" is:
   a. The document that specifies the order (priority) in which project documents will be used when it becomes necessary to resolve inconsistencies between project documents
   b. The order in which project tasks should be completed
   c. The relationship that project tasks have to one another
   d. The ordered list (by quality) of the screened vendors for a project deliverable

4. In which type of contract arrangement is the *contractor* least likely to want to control costs?
   a. Cost plus percentage of cost
   b. Firm-Fixed Price
   c. Time and Materials
   d. Purchase order

5. In which type of contract arrangement is the *contractor* most likely to want to control costs?
   a. Cost plus percentage of cost
   b. Firm-Fixed Price
   c. Time and Materials
   d. Fixed-Price-Incentive-Fee

6. In which type of contract arrangement is the *contractor* at the most risk of absorbing all cost overruns?
   a. Cost plus percentage of cost
   b. Firm-Fixed Price
   c. Time and Materials
   d. Cost-Plus-Incentive-Fee

7. In which type of contract arrangement is the *customer* at the most risk of absorbing excessive cost overruns?
   a. Cost plus percentage of cost
   b. Firm-Fixed Price
   c. Time and Materials
   d. Fixed-Price-Incentive-Fee

8. What is the primary objective the customer's project manager focuses on when selecting a contract type?
   a. Transferring all risk to the contractor
   b. Creating reasonable contractor risk with provisions for efficient and economical performance incentives for the contractor
   c. Retaining all project risk, therefore reducing project contract costs
   d. None of the above

9. Which type of contract arrangement is specifically designed to give a contractor relief for inflation or material/labor cost increases on a long-term contract?
   a. Cost plus percentage of cost
   b. Firm-Fixed Price
   c. Time and Materials
   d. Firm-Fixed Price with economic price adjustment

10. Which of the following is not a factor to consider when selecting a contract type?
   a. The type/complexity of the requirement
   b. The urgency of the requirement
   c. The extent of price competition
   d. All are factors to consider

11. In a Fixed-Price-Incentive-Fee contract, the "point of total assumption" refers to the point in the project cost curve where:
   a. The customer assumes responsibility for every additional dollar that is spent in fulfillment of the contract
   b. The contractor assumes responsibility for every additional dollar that is spent in fulfillment of the contract
   c. The price ceiling is reached after the contractor recovers the target profit
   d. None of the above

12. A written *preliminary* contractual instrument prepared prior to the issuance of a definitive contract that authorizes the contractor to begin work immediately, within certain limitations, is known as a:
   a. Definitive contract
   b. Preliminary contract
   c. Letter contract/Letter of intent
   d. Purchase order

13. A contract entered into after following normal procedures (i.e., negotiation of terms, conditions, cost, and schedule) but prior to initiation of performance is known as a:
   a. Definitive contract
   b. Completed contract
   c. Letter contract/letter of intent
   d. Pricing arrangement

14. Which of the following is not a function of the contract administration activity?
   a. Contract change management
   b. Specification interpretation
   c. Determination of contract breach
   d. Selection of the project manager

15. A fixed-price contract is typically sought by the project manager from the customer's organization when:

a. The risk and consequences associated with the contracted task are large and the customer wishes to transfer the risk

b. The project manager's company is proficient at dealing with the contracted activities

c. Neither the contractor nor the project manager understand the scope of the task

d. The project manager's company has excess production capacity

16. Which of the following is/are a typical action(s) a customer would take if the customer received nonconforming materials or products and the customer did not have the ability to bring the goods into conformance?

a. Reject the entire shipment but pay the full cost of the contract

b. Accept the entire shipment, no questions asked

c. Accept the shipment on condition that the nonconforming products will be brought into conformance by the vendor at the vendor's expense

d. Accept the shipment and resell it to a competitor

17. If a project manager requires the use of a piece of equipment, what is the breakeven point where leasing and renting are the same?

Cost Categories	Renting Costs	Leasing Costs
Annual Maintenance	$ 0.00	$3,000.00
Daily Operation	$ 0.00	$ 70.00
Daily Rental	$100.00	$ 0.00

a. 300 days

b. 30 days

c. 100 days

d. 700 days

18. In which type of incentive contract is there a maximum or minimum value established on the profits allowed for the contract?

a. Cost-Plus-Incentive-Fee contract

b. Fixed-Price-Incentive-Fee contract

c. Time and Material Incentive Fee contract

d. Split Pricing Incentive Fee contract

19. In which type of incentive contract is there a maximum or minimum value established on the final price of the contract?

a. Cost-Plus-Incentive-Fee contract

b. Fixed-Price-Incentive-Fee contract

c. Time and Material Incentive Fee contract

d. Split-Pricing-Incentive-Fee contract

Answers to multiple-choice questions:

1. c
2. a
3. a
4. a
5. b
6. b
7. a
8. b

9. d
10. d
11. b
12. c
13. a
14. d
15. a
16. c
17. c
18. a
19. b

The following multiple-choice questions will be helpful in reviewing the principles of **Professional and Social Responsibility**:

1. You have been sent on a business trip to visit one of the companies bidding on a contract to be awarded by your company. You are there to determine the validity of the information in their proposal. They take you to dinner one evening at a very expensive restaurant. When the bill comes, you should:
   a. Thank them for their generosity and let them pay the bill
   b. Thank them for their generosity and tell them that you prefer to pay for your own meal
   c. Offer to pay for the meal for everyone and put it on your company's credit card
   d. Offer to pay the bill, put it on your company's credit card, and make the appropriate adjustment in their bid price to cover the cost of the meals

2. You are preparing a proposal in response to a Request for Proposal (RFP) from a potentially important client. The salesperson in your company working on the proposal tells you to "lie" in the proposal to improve the company's chance of winning the contract. You should:
   a. Do as you are told
   b. Refuse to work on the proposal
   c. Report the matter to either your superior, the project sponsor, or the corporate legal group
   d. Resign from the company

3. You are preparing for a customer interface meeting and your project sponsor asks you to lie to the customer about certain test results. You should:
   a. Do as you are told
   b. Refuse to work on the project from this point forth
   c. Report the matter to either your superior or the corporate legal group for advice
   d. Resign from the company

4. One of the project managers in your company approaches you with a request to use some of the charge numbers from your project (which is currently running under budget) for work on their project (which is currently running over budget). Your contract is a cost-reimbursable contract for a client external to your company. You should:
   a. Do as you are requested
   b. Refuse to do this unless he allows you to use his charges number later on
   c. Report the matter to either your superior, the project sponsor, or the corporate legal group
   d. Ask the project manager to resign from the company

5. You have submitted a proposal to a client as part of a competitive bidding effort. One of the people evaluating your bid informs you that it is customary for you to send them some gifts in order to have a better chance of winning the contract. You should:
   a. Send them some gifts
   b. Do not send any gifts and see what happens
   c. Report the matter to either your superior, the project sponsor, or the corporate legal group for advice
   d. Withdraw the proposal

6. You just discovered that the company in which your brother-in-law is employed has submitted a proposal to your company. Your brother-in-law has asked you to do everything possible to make sure that his company will win the contract because his job may be in jeopardy. You should:
   a. Do what your brother-in-law requests
   b. Refuse to look into the matter and pretend it never happened
   c. Report the conflict of interest to either your superior, the project sponsor, or the corporate legal group
   d. Hire an attorney for advice

7. As part of a proposal evaluation team, you have discovered that the contract will be awarded to Alpha Company and that a formal announcement will be made in two days. The price of Alpha Company's stock may just skyrocket because of this contract award. You should:
   a. Purchase as much Alpha Company stock as you can within the next two days
   b. Tell family members to purchase the stock
   c. Tell employees in the company to purchase the stock
   d. Do nothing about stock purchases until after the formal announcement has been made

8. Your company has decided to cancel a contract with Beta Company. Only a handful of employees know about this upcoming cancellation. The announcement of the cancellation will not be made for about two days from now. You own several shares of Beta Company stock and know full well that the stock will plunge on the bad news. You should:
   a. Sell your stock as quickly as possible
   b. Sell your stock and tell others whom you know own the stock to do the same thing
   c. Tell the executives to sell their shares if they are stock owners
   d. Do nothing until after the formal announcement is made

9. You are performing a two-day quality audit of one of your suppliers. The supplier asks you to remain a few more days so that they can take you out deep sea fishing and gambling at the local casino. You should:
   a. Accept as long as you complete the audit within two days
   b. Accept but take vacation time for fishing and gambling
   c. Accept their invitation but at a later time so that it does not interfere with the audit
   d. Gracefully decline their invitation

10. You have been assigned as the project manager for a large project in the Pacific Rim. This is a very important project for both your company and the client. In your first meeting with the client, you are presented with a very expensive gift for yourself and another expensive gift for your wife. You were told by your company that this is considered an acceptable custom when doing work in this country. You should:
   a. Gracefully accept both gifts
   b. Gracefully accept both gifts but report only your gift to your company
   c. Gracefully accept both gifts and report both gifts to your company
   d. Gracefully refuse the acceptance of either gift

11. Your company is looking at the purchase of some property for a new plant. You are part of the committee making the final decision. You discover that the owner of a local auto dealership from whom you purchase family cars owns one of the properties. The owner of the dealership tells you in confidence that he will give you a new model car to use for free for up to three years if your company purchases his property for the new plant. You should:
   a. Say thank you and accept the offer
   b. Remove yourself from the committee for conflict of interest
   c. Report the matter to either your superior, the project sponsor, or the corporate legal group for advice
   d. Accept the offer as long as the car is in your spouse's name

12. Your company has embarked upon a large project (with you as project manager) and as an output from the project there will be some toxic waste as residue from the manufacturing operations. A subsidiary plan has been developed for the containment and removal of the toxic waste and no environmental danger exists. This information on toxic waste has not been made available to the general public as yet, and the general public does not appear to know about this waste problem. During an interview with local newspaper personnel you are discussing the new project and the question of environmental concerns comes up. You should:
   a. Say there are no problems with environmental concerns
   b. Say that you have not looked at the environmental issues problems as yet
   c. Say nothing and ask for the next question
   d. Be truthful and reply as delicately as possible

13. As a project manager, you establish a project policy that you, in advance of the meeting, review all handouts presented to your external customer during project status review meetings. While reviewing the handouts, you notice that one slide contains company confidential information. Presenting this information to the customer would certainly enhance good will. You should:
   a. Present the information to the customer
   b. Remove the confidential information immediately
   c. Discuss the possible violation with senior management and the legal department before taking any action
   d. Discuss the situation first with the team member that created the slide, and then discuss the possible violation with senior management and the legal department before taking any action

14. You are managing a project for an external client. Your company developed a new testing procedure to validate certain properties of a product and the new testing procedure was developed entirely with internal funds. Your company owns all of the intellectual property rights associated with the new test. The workers that developed the new test used one of the components developed for your current customer as part of the experimental process. The results using the new test showed that the component would actually exceed the customer's expectations. You should:
   a. Show the results to the customer but do not discuss the fact that it came from the new test procedure
   b. Do not show the results of the new test procedure since the customer's specifications call for use of the old test procedures
   c. Change the customer's specifications first and then show the customer the results
   d. Discuss the release of this information with your legal department and senior management before taking any action

15. Using the same scenario as in the previous question, assume that the new test procedure that is expected to be more accurate than the old test procedure indicates that performance will not meet customer specifications whereas the old test indicates that customer specifications will be barely met. You should:
   a. Present the old test results to the customer showing that specification requirements will be met
   b. Show both sets of test results and explain that the new procedure is unproven technology
   c. Change the customer's specifications first and then show the customer the results
   d. Discuss the release of this information with your legal department and senior management before taking any action

16. Your customer has demanded to see the "raw data" test results from last week's testing. Usually the test results are not released to customers until after the company reaches a conclusion on the meaning of the test results. Your customer has heard from the grapevine that the testing showed poor results. Management has left the entire decision up to you. You should:
    a. Show the results and explain that it is simply raw data and that your company's interpretation of the results will be forthcoming
    b. Withhold the information until after the results are verified
    c. Stall for time even if it means lying to the customer
    d. Explain to the customer your company's policy of not releasing raw data

17. One of your team members plays golf with your external customer's project manager. You discover that the employee has been feeding the customer company-sensitive information. You should:
    a. Inform the customer that project information from anyone other than the project manager is not official until released by the project manager
    b. Change the contractual terms and conditions and release the information
    c. Remove the employee from your project team
    d. Explain to the employee the ramifications of his actions and that he still represents the company when not at work, then report this as a violation

18. Your company has a policy that all company-sensitive material must be stored in locked filing cabinets at the end of each day. One of your employees has received several notices from the security office for violating this policy. You should:
    a. Reprimand the employee
    b. Remove the employee from your project
    c. Ask the Human Resources Group to have the employee terminated
    d. Counsel the employee as well as other team members on the importance of confidentiality and the possible consequences for violations

19. You have just received last month's earned value information that must be shown to the customer in the monthly status review meeting. Last month's data showed unfavorable variances that exceeded the permissible threshold limits on time and cost variances. This was the result of a prolonged power outage in the manufacturing area. Your manufacturing engineer tells you that this is not a problem and next month you will be right on target on time and cost as you have been in the last five months. You should:
    a. Provide the data to the customer and be truthful in the explanation of the variances
    b. Adjust the variances so that they fall within the threshold limits since this problem will correct itself next month
    c. Do not report any variances this month
    d. Expand the threshold limits on the acceptable variances but do not tell the customer

20. You are working in a foreign country where it is customary for a customer to present gifts to the contractor's project manager throughout the project as a way of showing appreciation. Declining the gifts would be perceived by the customer as an insult. Your company has a policy on how to report gifts received. The *best* way to handle this situation would be to:
    a. Refuse all gifts
    b. Send the customer a copy of our company's policy on accepting gifts
    c. Accept the gifts and report the gifts according to policy
    d. Report all gifts even though the policy says that some gifts need not be reported

21. You are interviewing a candidate to fill a project management position in your company. On his resume, he states that he is a PMP®. One of your workers who knows the candidate informs you that he is not a PMP® yet but is planning to take the test next month and certainly expects to pass. You should:
   a. Wait until he passes the exam before interviewing him
   b. Interview him and ask him why he lied
   c. Inform PMI® of the violation
   d. Forget about it and hire him if he looks like the right person for the job

22. You are managing a multinational project from your office in Chicago. Half of your project team are from a foreign country but are living in Chicago while working on your project. These people inform you that two days during next week are national religious holidays in their country and they will be observing the holiday by not coming into work. You should:
   a. Respect their beliefs and say nothing
   b. Force them to work because they are in the United States where their holiday is not celebrated
   c. Tell them that they must work noncompensated overtime when they return to work in order to make up the lost time
   d. Remove them from the project team if possible

23. PMI® informs you that one of your team members who took the PMP® exam last week and passed may have had the answers to the questions in advance provided to him by some of your other team members who are also PMP®s and were tutoring him. PMI® is asking for your support in the investigation. You should:
   a. Assist PMI® in the investigation of the violation
   b. Call in the employee for interrogation and counseling
   c. Call in the other team members for interrogation and counseling
   d. Tell PMI® that it is their problem, not your problem

24. One of your team members has been with you for the past year since his graduation from college. The team member informs you that he is now a PMP® and shows you his certificate from PMI® acknowledging this. You wonder how he was qualified to take the exam since he had no prior work experience prior to joining your company one year ago. You should:
   a. Report this to PMI® as a possible violation
   b. Call in the employee for counseling
   c. Ask the employee to surrender his PMP® credentials
   d. Do nothing

25. Four companies have responded to your RFP. Each proposal has a different technical solution to your problem and each proposal states that the information in the proposal is company-proprietary knowledge and not to be shared with anyone. After evaluation of the proposals, you discover that the best technical approach is from the highest bidder. You are unhappy about this. You decide to show the proposal from the highest bidder to the lowest bidder to see if the lowest bidder can provide the same technical solution but at a lower cost. This situation is:
   a. Acceptable since once the proposals are submitted to your company, you have unlimited access to the intellectual property in the proposals
   b. Acceptable since all companies do this
   c. Acceptable as long as you inform the high bidder that you are showing their proposal to the lowest bidder
   d. Unacceptable and is a violation of the Code of Professional Conduct

Answers to the multiple-choice questions:

1. b
2. c
3. c
4. c
5. c
6. c
7. d
8. d
9. d
10. c
11. c
12. d
13. d
14. d
15. d
16. a
17. d
18. d
19. a
20. d
21. c
22. a
23. a
24. a
25. d

Chapter Twenty-Five

PROJECT MANAGEMENT LOGIC PROBLEMS

Logic problems are a unique way of learning how to apply project management principles to project situations. They are enjoyable and entertaining. Include these problems in your project management studies to enhance the learning experience.

How to Solve Logic Problems[1]

To solve logic problems, the wording in the clues must be read carefully. All of the necessary information is provided in the clues, and trial-and-error solutions are not necessary. The problems can be solved with simple logic. In some cases, grids are provided. Shown below is the technique for solving the grid problems.

Sample Logic Problem:

Three married couples live in three different cities and each couple has a different number of children. The husbands are Alfred, Dirk, and Mickey. The wives are Annette, Barbara, and Cathy but not necessarily married to the husbands according to the order of the husbands' names. The three cities in which the couples live are Boston, Chicago, and Denver, and the numbers of children are one, two, and three. In completing the table below, we will use a "•" to signify "yes" and an "X" to signify "No."

Clues:

1. Barbara and Mickey are a couple but do not live in Chicago.
2. One of the couples which does not include Annette live in Boston with their three children.
3. Alfred and his wife live in Denver; they have more than one child.
4. Cathy's husband's first name begins with an initial that appears later in the alphabet than Cathy's first name initial.

---

[1] For additional problems, see Kerzner, H., *Logic Problems for Project Managers*, John Wiley & Sons Publishers, 2006.

Explanation:

From clue 1, Barbara and Mickey are husband and wife. So, a "●" can be placed at the intersection and the other cells have an "X." Likewise, because they do not live in Chicago, we can place an "X" in that location as well, and any other locations that relate Barbara or Mickey to Chicago. This is shown in the figure below.

		Husband			City			Children		
		Alfred	Dirk	Mickey	Boston	Chicago	Denver	1	2	3
Wife	Annette			X						
	Barbara	X	X	●		X				
	Cathy			X						
Children	1									
	2									
	3									
City	Boston									
	Chicago			X						
	Denver									

With clue 2, we can signify a "Yes" in the box that signifies the intersection of Boston and three children, and also a "No" for the Annette location in that column. This is shown below.

		Husband			City			Children		
		Alfred	Dirk	Mickey	Boston	Chicago	Denver	1	2	3
Wife	Annette			X	X					
	Barbara	X	X	●		X				
	Cathy			X						
Children	1				X					
	2				X					
	3				●	X	X			
City	Boston									
	Chicago			X						
	Denver									

Using clue 3, we can place a "Yes" in the intersection of Alfred and Denver, and the corresponding cells will then have a "No" response. We can also eliminate the one child cell from the Alfred row and the Denver column. The result is shown below.

		Husband			City			Children		
		Alfred	Dirk	Mickey	Boston	Chicago	Denver	1	2	3
Wife	Annette			X	X					
	Barbara	X	X	●		X				
	Cathy			X						
Children	1	X			X					
	2				X					
	3				●	X	X			
City	Boston	X								
	Chicago	X		X						
	Denver	●	X	X						

Now notice that in the cells that relate Husbands and Cities, we know that Dirk must reside in Chicago and Mickey must reside in Boston. Also, Mickey must live in Boston with three children. The results are shown below.

		Husband			City			Children		
		Alfred	Dirk	Mickey	Boston	Chicago	Denver	1	2	3
Wife	Annette			X	X					X
	Barbara	X	X	●	●	X	X	X	X	●
	Cathy			X	X					X
Children	1	X	●	X	X	●	X			
	2	●	X	X	X	X	●			
	3	X	X	●	●	X	X			
City	Boston	X	X	●						
	Chicago	X	●	X						
	Denver	●	X	X						

We also know from this figure that the city in which the couple has two children is Denver. This is represented in the figure below.

		Husband			City			Children		
		Alfred	Dirk	Mickey	Boston	Chicago	Denver	1	2	3
Wife	Annette			X	X					X
	Barbara	X	X	●	●	X	X	X	X	●
	Cathy			X	X					X
Children	1	X	●	X	X	●	X			
	2	●	X	X	X	X	●			
	3	X	X	●	●	X	X			
City	Boston	X	X	●						
	Chicago	X	●	X						
	Denver	●	X	X						

Now using clue 4, we know that Cathy cannot be married to Alfred. Therefore, we can complete the figure as shown below.

		Husband			City			Children		
		Alfred	Dirk	Mickey	Boston	Chicago	Denver	1	2	3
Wife	Annette	●	X	X	X	X	●	X	●	X
	Barbara	X	X	●	●	X	X	X	X	●
	Cathy	X	●	X	X	●	X	●	X	X
Children	1	X	●	X	X	●	X			
	2	●	X	X	X	X	●			
	3	X	X	●	●	X	X			
City	Boston	X	X	●						
	Chicago	X	●	X						
	Denver	●	X	X						

We now know the solution to the problem:

- Alfred and Annette live in Denver with two children.
- Mickey and Barbara live in Boston with three children.
- Dirk and Cathy live in Chicago with one child.

Sometimes, you may have to go back over the clues a second or third time to extract more information. However, guessing is never required and logic must prevail.

For other types of logic problems, the difficulty is in knowing which clue or clues to start with. Since this may be more complex, hints are provided to assist you.

Now it's your turn to demonstrate your decision-making skills. Good luck!

Logic Problem #1: Types of Contracts

In the figure below are six different contract types for six different project managers. From the clues under the figure, determine the contract type for each project manager and their position in the figure. Knowledge of each type of contract is essential to solve the problem.

Project Manager: Jane, Richard, Paul, Frank, Tim, Alice

Contract Type: Cost-Plus-Percentage-of-Cost (CPPC), Firm-Fixed-Price (FFP), Cost-Plus-Incentive-Fee (CPIF), Fixed-Price-Incentive-Fee (FPIF), Cost-Plus-Fixed-Fee (CPFF), Time & Materials (T & M)

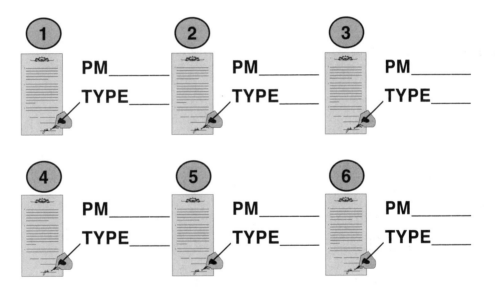

Clues:

1. The contract type, which today is illegal in the government, is located in position 4 in the diagram.
2. Tim's project has a contract that contains a "ceiling" and "floor" on profits.
3. Alice's contract is in an even-numbered position in the diagram.
4. The two incentive-type contracts are located in different rows in the figure and horizontally adjacent to only one other contract, and both of the adjacent contracts are for projects managed by female project managers.
5. Frank's project, which is not in position 3, and Richard's project are not adjacent to each other either horizontally or vertically, and are located in different rows.

6. The contract type with the maximum risk exposure to the seller is to the right (as you look at the figure) and next to the Cost-Plus contract where the fee is defined as a dollar value rather than as a percentage, and above the contract type that contains a "Point of Total Assumption" term.

Logic Problem #2: The Mysterious Network Diagram

A project manager discovers that his team has neglected to complete the network diagram for the project. The network diagram is shown below. However, the project manager has some information available, specifically that each activity, labeled A–G, has a different duration between one and seven weeks. Also, the slack time for each of the activities is known as shown below.

Duration (weeks):   1, 2, 3, 4, 5, 6, 7

Slack time (weeks): 0, 0, 0, 2, 4, 4, 7

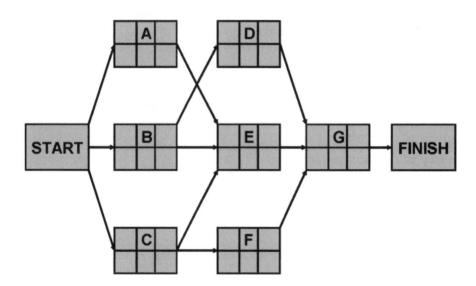

Using the clues provided below, determine the duration of each activity as well as the early start, early finish, latest start, and latest finish times for each activity.

Clues:

1. Activity E is on the critical path.
2. The early start time (ES) for Activity F is five weeks.
3. The duration of Activity B is seven weeks.
4. Activity D has four weeks of slack, but Activity F has the greatest amount of slack.
5. The early finish time (EF) for Activity G is seventeen weeks.
6. The latest finish time (LF) for Activity E is thirteen weeks.

Activity	Duration	Early Start	Early Finish	Latest Start	Latest Finish
A	___	___	___	___	___
B	___	___	___	___	___
C	___	___	___	___	___
D	___	___	___	___	___
E	___	___	___	___	___
F	___	___	___	___	___
G	___	___	___	___	___

Logic Problem #3: The Incomplete Status Report

As a continuation of Logic Problem #2, you also discover that the status report for your project is somewhat incomplete. The status report is shown below. From the clues provided, complete the status report.

Clues:

1. For Activity B, PV = $100
2. For Activity C, EV = $200
3. For Activity D, AC = $300
4. Activity A has not started yet.
5. For all of the activities EV (Total) = $930

Activity	PV	EV	AC	SV	CV
A	___	___	___	($100)	0
B	___	___	___	50	60
C	___	___	___	60	60
D	___	___	___	100	80
E	___	___	___	120	(20)
Total	___	___	___	___	___

Logic Problem #4: Another Mysterious Network Diagram

A project manager discovers that his team has neglected to complete the network diagram for the project. The network diagram is shown below. However, the project manager has some information available, specifically that each activity, labeled A–G, has a different duration between one and seven weeks. Also, the slack time for each of the activities is known as shown below.

Duration (weeks):   1, 2, 3, 4, 5, 6, 7

Slack time (weeks):   0, 0, 0, 1, 1, 3, 7

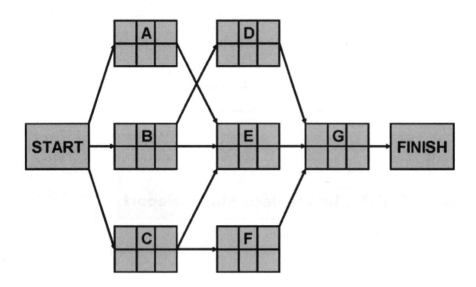

Using the clues provided below, determine the duration of each activity as well as the early start, early finish, latest start, and latest finish times for each activity.

Clues:

1. Activity E is the longest duration activity and is on the critical path, which is the unlucky number thirteen; also, there is only one critical path.
2. The early finish time (EF) for Activity F is eleven weeks.
3. The latest start time (LS) for Activity D is nine weeks.
4. If Activity A slips by one week, it will be on a critical path.

Activity	Duration	Early Start	Early Finish	Latest Start	Latest Finish
A	_____	_____	_____	_____	_____
B	_____	_____	_____	_____	_____
C	_____	_____	_____	_____	_____
D	_____	_____	_____	_____	_____
E	_____	_____	_____	_____	_____
F	_____	_____	_____	_____	_____
G	_____	_____	_____	_____	_____

Logic Problem #5: Another Mysterious Network Diagram

A project manager discovers that his team has neglected to complete the network diagram for the project. The network diagram is shown below. However, the project manager has some information available, specifically that each activity, labeled A–G, has a different duration between one and seven weeks. Also, the slack time for each of the activities is known as shown on the next page.

Duration (weeks):   1, 2, 3, 4, 5, 6, 7

Slack time (weeks): 0, 0, 0, 3, 6, 8, 8

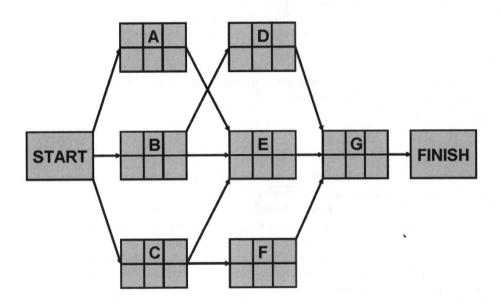

Using the clues provided below, determine the duration of each activity as well as the early start, early finish, latest start, and latest finish times for each activity.

Clues:

1. There exists only one critical path and it is the largest possible number given the possible durations shown.
2. Activity E has the smallest amount of slack that is greater than zero.
3. The early finish time (EF) for Activity A is four weeks and this does not equal the latest finish time (LF). (Note: There is no negative slack in the network.)
4. The slack in Activity C is eight weeks.
5. The duration of Activity F is greater than the duration of Activity C by at least two weeks.

Activity	Duration	Early Start	Early Finish	Latest Start	Latest Finish
A	_____	_____	_____	_____	_____
B	_____	_____	_____	_____	_____
C	_____	_____	_____	_____	_____
D	_____	_____	_____	_____	_____
E	_____	_____	_____	_____	_____
F	_____	_____	_____	_____	_____
G	_____	_____	_____	_____	_____

Logic Problem #6: Integration Management (*PMBOK® Guide* 2004)

Five project managers were assigned the responsibility of teaching the first five sections of the Integration Management Chapter in the *PMBOK® Guide* 2004. Each project manager taught a different process, required a different amount of time for teaching, and created a different number of multiple-choice questions as a review. Using the clues and the figure below, determine which project manager taught each process, the time required, and the number of review questions.

Note: Knowledge of this chapter in the *PMBOK® Guide* 2004 is required to do this problem. Although there are seven sections to Integration Management, do not consider the last two sections of the Integration Management Chapter when doing this problem.

Clues:

1. Maggie taught one of the two processes that describe four tools and techniques; she required two hours to teach the subject and used an exam with twelve questions.
2. The lecture of 2.5 hours was for a process where the project charter was an input.
3. Five test questions and eight test questions were used in the two processes where either the statement of work or the scope statement was not an input, although not necessarily in this order.
4. Quinton taught for 2.5 hours on the only process where the scope statement was an output.
5. Alfonso taught the only process where expert judgment was not part of the tools and techniques, and used three hours to teach the material.
6. Susan taught one of the two processes where enterprise environmental factors were not an input.
7. The eight-question test accompanied a 1.5-hour lecture for a process where organizational process assets were not an input; this was not the section on process execution.
8. Ten questions were not used in the process involving monitoring and control.

	Alfonso	Dirk	Maggie	Quinton	Susan	5	8	10	12	15	1	1.5	2	2.5	3
Develop Charter															
Scope Statement															
Mgt. Plan															
Execution															
Monitor & Control															
1															
1.5															
2															
2.5															
3															
5															
8															
10															
12															
15															

Name — Test Questions — Time (hrs) (column group headers)

Process (row group label for Develop Charter through Monitor & Control)
Time (hrs) (row group label for 1 through 3)
Test Questions (row group label for 5 through 15)

PM	Process	No. of Questions	Time (hrs)

Logic Problem #7: Scope Management (*PMBOK® Guide* 2004)

Five project managers were assigned the responsibility of teaching the five sections of the Scope Management Chapter in the *PMBOK® Guide* 2004. Each project manager taught a different process, required a different amount of time for teaching, and created a different number of multiple-choice questions as a review. Using the clues and the figure below, determine which project manager taught each process, the time required, and the number of review questions.

Note: Knowledge of this chapter in the *PMBOK® Guide* 2004 is required to do this problem. Also, the names of the people in this problem and the next several problems related to the *PMBOK® Guide* 2004 are the same, but the first name and last name may not be the same from problem to problem.

Clues:

1. Susan needed two hours to teach the process where the work breakdown structure dictionary was an output and Dirk used nine questions for the process where the work breakdown dictionary (updates) was an output.
2. Alfonso taught one of the modules where expert judgment was a tool and technique.
3. The process where stakeholder analysis was a tool and technique had six test questions, one for each hour of lecture.
4. Maggie used eight test questions and five hours to teach the only process where the scope management plan was not an output.
5. Three hours were required to teach the module where replanning was a tool and technique.
6. The exam with five questions was not with the shortest lecture.

		Name					Test Questions					Time (hrs)				
		Alfonso	Dirk	Maggie	Quinton	Susan	5	6	7	8	9	2	3	4	5	6
Process	Scope Planning															
	Scope Definition															
	Create WBS															
	Verification															
	Scope Control															
Time (hrs)	2															
	3															
	4															
	5															
	6															
Test Questions	5															
	6															
	7															
	8															
	9															

PM	Process	No. of Questions	Time (hrs)

Logic Problem #8: Time Management (*PMBOK® Guide* 2004)

Five project managers were assigned the responsibility of teaching the first five sections of the Time Management Chapter in the *PMBOK® Guide* 2004. Each project manager taught a different process, required a different amount of time for teaching, and created a different number of multiple-choice questions as a review. Using the clues and the figure below, determine which project manager taught each process, the time required, and the number of review questions.

Note: Knowledge of this chapter in the *PMBOK® Guide* 2004 is required to do this problem. Consider only the first five processes in this chapter, although there are more.

Clues:

1. The largest number of test questions went with the longest lecture.
2. The process where the WBS dictionary was an input used up all of Dirk's five hours of lecture.
3. The process with the greatest number of tools and techniques was taught in six hours and with only eight review questions.
4. The process with the least number of outputs was taught by Quinton using twelve test questions, but not in three hours.
5. Maggie used three hours to teach one of the processes where PM software was a tool and technique, but did not use fifteen test questions.

6. Susan did not teach the process where organizational process assets were an input.

Logic Problem #9: Cost Management (*PMBOK® Guide* 2004)

Three project managers were assigned the responsibility of teaching the three sections of the Cost Management Chapter in the *PMBOK® Guide* 2004. Each project manager taught a different process, required a different amount of time for teaching, and created a different number of multiple-choice questions as a review. Using the clues and the figure below, determine which project manager taught each process, the time required, and the number of review questions.

Note: Knowledge of this chapter in the *PMBOK® Guide* 2004 is required to do this problem.

Clues:

1. Dirk taught the module with the largest number of outputs.
2. Twelve hours were required to teach the module with the largest number of inputs.
3. Nine hours of lectures were used with the only module where PM software was not a tool.
4. Maggie lectured for six hours.

		Name			Test Questions			Time (hrs)		
		Alfonso	Dirk	Maggie	10	15	20	6	9	12
Process	Estimating									
	Budgeting									
	Control									
Time (hrs)	6									
	9									
	12									
Test Questions	10									
	15									
	20									

PM	Process	No. of Questions	Time (hrs)

Logic Problem #10: Risk Management (*PMBOK*® *Guide* 2004)

Five project managers were assigned the responsibility of teaching the first five sections of the Risk Management Chapter in the *PMBOK*® *Guide* 2004. Each project manager taught a different process, required a different amount of time for teaching, and created a different number of multiple-choice questions as a review. Using the clues and the figure below, determine which project manager taught each process, the time required, and the number of review questions.

Note: Knowledge of this chapter in the *PMBOK*® *Guide* 2004 is required to do this problem. Consider only the first five processes in this chapter although there are more.

Clues:

1. Dirk taught one of the three processes where the risk register (update) was an output, but did not use twelve hours or thirty questions.
2. Maggie lectured for sixteen hours on the only module with project scope management not an input.
3. Assumption analysis is a tool and technique in the process that requires eight hours for Susan to teach.
4. Quinton teaches one of the two modules where the project management plan is an input; his class does not have fifteen questions.
5. The process with the documentation review as a tool has twenty-five test questions.

6. Alfonso used fifteen test questions to go with ten hours of lectures.
7. The only process with more than one output uses twenty questions.

		Surname					Test Questions					Time (hrs)				
		Alfonso	Dirk	Maggie	Quinton	Susan	10	15	20	25	30	8	10	12	14	16
Process	Risk Planning															
	Risk Identification															
	Qual. Analysis															
	Quant. Analysis															
	Response Planning															
Time (hrs)	8															
	10															
	12															
	14															
	16															
Test Questions	10															
	15															
	20															
	25															
	30															

PM	Process	No. of Questions	Time (hrs)

Logic Problem #11: Procurement Management (*PMBOK® Guide* 2004)

Five project managers were assigned the responsibility of teaching the first five sections of the Procurement Management Chapter in the *PMBOK® Guide* 2004. Each project manager taught a different process, required a different amount of time for teaching, and created a different number of multiple-choice questions as a review. Using the clues and the figure below, determine which project manager taught each process, the time required, and the number of review questions.

Note: Knowledge of this chapter in the *PMBOK® Guide* 2004 is required to do this problem. Consider only the first five processes in this chapter although there are more.

Clues:

1. The make-or-buy analysis required eight hours of lecture but not by Quinton.
2. The process where the make-or-buy decision is an output required ten hours of lecture and thirty questions.
3. Dirk discussed bidder conferences in his lectures and five of his twenty-five questions were on this subject.
4. Maggie lectured on the contract evaluation criteria as part of her six-hour lecture.

5. The three-hour lecture was not accompanied by twenty-five test questions; this was not the contract administration process that was covered by Susan's lectures.
6. Vendor payment systems, as a tool and technique, have a ten-question test.
7. The discussion of the contract types, as a tool and technique, as expected has the test with the greatest number of questions.

Logic Problem #12: Human Resource Management (*PMBOK® Guide* 2004)

Four project managers were assigned the responsibility of teaching the four sections of the Human Resource Management Chapter in the *PMBOK® Guide* 2004. Each project manager taught a different process, required a different amount of time for teaching, and created a different number of multiple-choice questions as a review. Using the clues and the figure below, determine which project manager taught each process, the time required, and the number of review questions.

Note: Knowledge of this chapter in the *PMBOK® Guide* 2004 is required to do this problem.

Clues:

1. Dirk prepared fifteen questions for the process with the greatest number of tools and techniques; this process did not involve the least or greatest number of teaching hours.

2. Alfonso spent six hours covering the process where roles and responsibilities were an output.
3. Maggie spent five hours on her module.
4. The module where virtual teams were discussed as tools and techniques had ten test questions.
5. Quinton used twelve test questions.

	Name				Test Questions				Time (hrs)			
	Alfonso	Dirk	Maggie	Quinton	10	12	14	15	3	4	5	6
HR Planning												
Acquire Team												
Develop Team												
Manage Team												
3												
4												
5												
6												
10												
12												
14												
15												

Process (rows HR Planning–Manage Team), Time (hrs) (rows 3–6), Test Questions (rows 10–15)

PM	Process	No. of Questions	Time (hrs)

Logic Problem #13: Communications Management (*PMBOK® Guide* 2004)

Four project managers were assigned the responsibility of teaching the four sections of the Communications Management Chapter in the *PMBOK® Guide* 2004. Each project manager taught a different process, required a different amount of time for teaching, and created a different number of multiple-choice questions as a review. Using the clues and the figure below, determine which project manager taught each process, the time required, and the number of review questions.

Note: Knowledge of this chapter in the *PMBOK® Guide* 2004 is required to do this problem.

Clues:

1. Dirk prepared ten test questions for the module he was teaching that had a communications management plan as an output; Maggie taught a process area where the communications management plan was an input.
2. The process where lessons learned were part of tools and techniques was taught in three hours.
3. Quinton used two hours to teach the process area where request changes was an output.
4. Communications methods as tools and techniques had six test questions and did not require five hours.
5. The twelve-question test was in the three-hour lecture; this was not Alfonso.

PM	Process	No. of Questions	Time (hrs)

Logic Problem #14: Quality Management (*PMBOK® Guide* 2004)

Three project managers were assigned the responsibility of teaching the three sections of the Communications Management Chapter in the *PMBOK® Guide* 2004. Each project manager taught a different process, required a different amount of time for teaching, and created a different number of multiple-choice questions as a review. Using the clues and the figure below, determine which project manager taught each process, the time required, and the number of review questions.

Note: Knowledge of this chapter in the *PMBOK® Guide* 2004 is required to do this problem.

Clues:

1. The place where design of experiments was part of tools and techniques was part of an eight-hour lecture. This was not Maggie.
2. Dirk lectured on Pareto Charts, which was clearly identified under tools and techniques.
3. The output of quality metrics was tested on as part of a thirty-question test.
4. Quality audits, as part of tools and techniques, was part of a ten-hour lecture and part of a test that had more than twenty questions.

		Name			Test Questions			Time (hrs)		
		Alfonso	Dirk	Maggie	20	25	30	8	10	12
Process	Planning									
Process	Assurance									
Process	Control									
Time (hrs)	8									
Time (hrs)	10									
Time (hrs)	12									
Test Questions	20									
Test Questions	25									
Test Questions	30									

PM	Process	No. of Questions	Time (hrs)

Logic Problem #15: The Earned Value Measurement Report

In the figure below is an earned value measurement report with fifteen open entries. In each of the entries, the numbers between $100 and $1,500 appear in increments of $100. Using the figure below and the clues, determine the location of each of the fifteen numbers.

Number: $100; $200; $300; $400; $500; $600; $700; $800; $900; $1,000; $1,100; $1,200; $1,300; $1,400; $1,500

Clues:

1. For activity D, the magnitude of CV is greater than the magnitude of SV.

2. Reading down in the EV column, there is a $900 immediately below the $600, and a $1,400 immediately below the $900.

3. Both activities A and B have favorable schedule variances but unfavorable cost variances; the cost variance for activity A is twice the cost variance for activity B.

4. Only one activity, which is not activity B, has an unfavorable schedule variance; the largest schedule variance in magnitude is $300; all of the cost variances fall between +/– $200.

5. Reading diagonally downward from left to right, there is a $100, $1,100, and $500 on one of the diagonals.

Activity	PV	EV	AC
A	____	____	____
B	____	____	____
C	____	____	____
D	____	____	____
E	____	____	____

Answers

Logic Problem #1

Position #1; Paul; CPIF
Position #2; Alice; CPFF
Position #3; Richard; FFP
Position #4; Frank; CPPC
Position #5; Jane; T & M
Position #6; Tim; FPIC

Logic Problem #2

Activity	Duration	ES	EF	LS	LF
A	3	0	3	4	7
B	7	0	7	0	7
C	5	0	5	2	7
D	2	7	9	11	13
E	6	7	13	7	13
F	1	5	6	12	13
G	4	13	17	13	17

Logic Problem #3

Activity	PV	EV	AC
A	100	0	0
B	100	150	90
C	140	200	140
D	280	380	300
E	80	200	220
	700	930	750

Logic Problem #4

Activity	Duration	ES	EF	LS	LF
A	4	0	4	1	5
B	2	0	2	3	5
C	5	0	5	0	5
D	3	2	5	9	12
E	7	5	12	5	12
F	6	5	11	6	12
G	1	12	13	12	13

The critical path is 13 weeks. Activity F has an early finish time of eleven weeks, but Activity F cannot be on the critical path because there is only one critical path and Activity E is on the critical path. Therefore, Activity G must have a duration of one week. We now know ES, EF, LS, and LF for Activity E. Because Activity F has an EF of eleven weeks, the durations of Activities C and F must be eleven weeks, and therefore five and six weeks. Activity C must be five weeks to align with the ES of five in Activity E. If Activity C were six weeks in duration, the length of the critical path could not be thirteen weeks. Using clue 3, and knowing that LF of Activity D must be twelve weeks, the duration of Activity D must be three weeks. Using clue 4, which tells us that Activity A has only one week of slack, and knowing that the LF for activity A is five weeks, the duration of Activity A must be four weeks.

Logic Problem #5

Activity	Duration	ES	EF	LS	LF
A	3	0	3	4	7
B	7	0	7	0	7
C	5	0	5	2	7
D	2	7	9	11	13
E	6	7	13	7	13
F	1	5	6	12	13
G	4	13	17	13	17

From the clues, Activities A, C, and E have slack. Activity F must also have slack since Activity C is on a path with it. Therefore, the critical path must be B-D-G. Since Activity C has eight weeks of slack, Activity F must also have eight weeks of slack which means, by elimination, that Activity A has six weeks of slack and the duration for Activity A must be four weeks. From clue 1, the critical path, which must have three and only three activities in it, must be the numbers five, six, and seven, and equal to eighteen weeks. From clue 5, the duration for Activity F must be three weeks and Activity C must be one week. Therefore, Activity E must be two weeks in duration. Now we can calculate ES, EF, LS, and LF for Activities C and F. Since the length of the critical path is eighteen weeks, we now know that the duration of Activity G must be six weeks. After calculating ES, EF, LS, and LF for Activity E, we know that the duration of Activity B must be seven weeks.

Logic Problem #6

Alfonso; Execution; 3; 5
Dirk; Management Plan; 1; 15
Maggie; Develop Charter; 2; 12
Quinton; Scope Statement; 2.5; 10
Susan; Monitor and Control; 1.5; 8

Logic Problem #7

Alfonso; Scope Definition; 6; 6
Dirk; Scope Control; 3; 9
Maggie; Scope Planning; 5; 8
Quinton; Verification; 4; 5
Susan; Create WBS; 2; 7

Logic Problem #8

Alfonso; Schedule Development; 6; 8
Dirk; Activity Definition; 5; 15
Maggie; Resource Estimating; 3; 10
Quinton; Duration Estimating; 4; 12
Susan; Sequencing; 7; 20

Logic Problem #9

Alfonso; Budgeting; 12; 15
Dirk; Control; 9; 20
Maggie; Estimating; 6; 10

Logic Problem #10

Alfonso; Risk Planning; 10; 15
Dirk; Quantitative Analysis; 14; 10
Maggie; Response Planning; 16; 20
Quinton; Qualitative Analysis; 12; 30
Susan; Risk Identification; 8; 25

Logic Problem #11

Alfonso; Plan Purchases; 8; 35
Dirk; Responses; 4; 25
Maggie; Select Sellers; 6; 20
Quinton; Plan Contracting; 10; 30
Susan; Contract Administration; 3; 10

Logic Problem #12

Alfonso; Human Resource Planning; 6; 14
Dirk; Develop Team; 4; 15
Maggie; Acquire Team; 5; 10
Quinton; Manage Team; 3; 12

Logic Problem #13

Alfonso; Manage Stakeholders; 4; 6
Dirk; Communications Planning; 5; 10
Maggie; Information Distribution; 3; 12
Quinton; Performance Reporting; 2; 8

Logic Problem #14

Alfonso; Planning; 8; 30
Dirk; Control; 12; 20
Maggie; Assurance; 10; 25

Logic Problem #15

Activity A; $100; $200; $400
Activity B; $1,000; $1,100; $1,200
Activity C; $300; $600; $500
Activity D; $800; $900; $700
Activity E; $1,500; $1,400; $1,300